OCR GCSE

Mathematics A

Foundation Homework Book

Series Editor: Brian Seager

- Howard Baxter
- Mike Handbury
- Jean Matthews
- Colin White
- Ruth Crookes
- John Jeskins
- Brian Seager

HODDER
EDUCATION
AN HACHETTE UK COMPANY

Hachette UK's policy is to use papers that are natural, renewable and recyclable products and made from wood grown in sustainable forests. The logging and manufacturing processes are expected to conform to the environmental regulations of the country of origin.

Orders: please contact Bookpoint Ltd, 130 Milton Park, Abingdon, Oxon OX14 4SB. Telephone: (44) 01235 827720. Fax: (44) 01235 400454. Lines are open from 9 a.m. to 6 p.m., Monday to Saturday, with a 24-hour message-answering service. Visit our website at www.hoddereducation.co.uk.

© Howard Baxter, Ruth Crookes, Michael Handbury, John Jeskins, Jean Matthews, Mark Patmore, Brian Seager, Colin White, Eddie Wilde, 2010
First published in 2010 by
Hodder Education
An Hachette UK company
338 Euston Road,
London, NW1 3BH

Impression number   5 4 3 2 1
Year   2015 2014 2013 2012 2011 2010

Cover illustration by Oxford Design & Illustrators
Typeset in 10pt Bembo by Pantek Arts Ltd, Maidstone, Kent
Printed by MPG Books, Bodmin

A catalogue record for this title is available from the British Library

ISBN: 978 1444 112 825

Contents

Introduction

About this book

This book contains exercises designed to be used for the Foundation tier of GCSE Mathematics. It is particularly aimed at the OCR Specification A and each exercise matches one in the OCR Foundation Student's Book.

In the Homework book, the corresponding exercises have the same number and end in H. Thus, for example, if you have been working on Coordinates from Unit A in class and used Exercise 7.1, then the homework exercise is 7.1H. The homework exercises cover the same mathematics.

As in the Student's Book, most of the questions in Unit B are designed to be done without a calculator so that you can practise for the non-calculator paper. Also, many of the questions require problem-solving skills. These are indicated by this icon.

These homework exercises provide extra practice and are also in a smaller book to carry home! If you have understood the topics, you should be able to tackle these exercises confidently as they are no harder than those you have done in class and in some cases may be a little easier. See if you agree. More practice helps to reinforce the ideas you have learned and makes it easier to remember at a later stage.

Brian Seager
Series Editor

Unit A Contents

Exercise 1.1H

1 Work out these.
    (a) 46 + 32   **(b)** 68 + 35   **(c)** 86 + 59
    (d) 178 + 26   **(e)** 185 + 232   **(f)** 188 + 346

2 Work out these.
    (a) 87 − 42   **(b)** 72 − 26   **(c)** 72 − 39
    (d) 239 − 87   **(e)** 273 − 148   **(f)** 634 − 276

3 Work out these.
    (a) 32 × 4   **(b)** 28 × 6   **(c)** 38 × 3
    (d) 49 × 8   **(e)** 135 × 7   **(f)** 278 × 5

4 Work out these.
    (a) 69 ÷ 3   **(b)** 54 ÷ 3   **(c)** 96 ÷ 6
    (d) 92 ÷ 4   **(e)** 165 ÷ 5   **(f)** 232 ÷ 4

5 Meena bought a top for £24, a pair of jeans for £37 and a belt for £9.
What was the total cost?

6 Karim earned £50.
He bought two CDs for £13 each.
How much did he have left?

7 Liz bought eight pens at 38p each.
What was the total cost in pence?

8 Will organises a disco.
The cost of hiring the hall and the DJ is £468.
How many £6 tickets does he need to sell to cover the cost?

Exercise 1.2H

1 List the following.
    (a) The multiples of 9 less than 100
    (b) The multiples of 12 less than 100

2 Use your answers to question **1** to list the common multiples of 9 and 12 less than 100.

3 Look at these numbers.
    4, 7, 9, 16, 18, 21, 25, 30, 42
    (a) Which have 2 as a factor?
    (b) Which have 3 as a factor?
    (c) Which have 7 as a factor?

4 List the following.
    (a) The multiples of 20 less than 130
    (b) The multiples of 25 less than 130

5 Use your answers to question **4** to find a common multiple of 20 and 25 less than 130.

6 List the following.
    (a) The factors of 20   **(b)** The factors of 36

7 Use your answers to question **6** to list the common factors of 20 and 36.

8 List the following.
    (a) The factors of 60   **(b)** The factors of 24

9 Use your answers to question **8** to list the common factors of 60 and 24.

10 Round these numbers to the nearest 1000.
    (a) 32 300   **(b)** 203 900   **(c)** 6600
    (d) 243 497   **(e)** 3 503 776

11 Round these numbers to the nearest 100.
    (a) 8392   **(b)** 21 830   **(c)** 354
    (d) 756 982   **(e)** 7032

12 Here are some newspaper headlines.
Round the numbers so that they have more impact.
    (a) Blues win with a majority of 7832!
    (b) £2 127 836 wasted by 'red tape'!
    (c) Moody goes to Real for €34 632 578!
    (d) Number passing goes up by 12 364!

Exercise 1.3H

1 Work out these.
(a) 71×10 (b) 84×100
(c) 26×1000 (d) 402×100
(e) $78 \times 10\,000$ (f) 80×100
(g) 617×1000 (h) 2800×100
(i) 140×1000 (j) $84 \times 10\,000$
(k) $974 \times 100\,000$ (l) $76 \times 1\,000\,000$

2 Work out these.
(a) $590 \div 10$ (b) $29\,000 \div 100$
(c) $648\,000 \div 1000$ (d) $92\,000 \div 100$
(e) $9\,200\,000 \div 1000$ (f) $789\,000 \div 10$
(g) $458\,200 \div 100$ (h) $8\,400\,000 \div 100$
(i) $71\,000\,000 \div 10\,000$

3 Work out these.
(a) 40×30 (b) 60×90
(c) 80×300 (d) 400×400
(e) 700×50 (f) 60×50
(g) 800×4000 (h) 500×200
(i) 900×8000 (j) 6000×2000
(k) $60\,000 \times 30$ (l) 7000×9000

4 Work out these.
(a) 53×20 (b) 63×40
(c) 165×50 (d) 73×400
(e) 82×700 (f) 59×600
(g) 537×800 (h) 96×3000

5 Work out these.
(a) 72×24 (b) 64×36
(c) 71×96 (d) 59×23
(e) 88×39 (f) 252×47
(g) 348×65 (h) 546×83
(i) 792×96 (j) 85×384

6 (a) How many centimetres are there in 589 metres?
(b) Change 4900 centimetres into metres.

7 1 hectare is $10\,000$ square metres.
How many square metres is 63 hectares?

8 Jeans cost £30 per pair.
What will eight pairs cost?

9 A sales representative drives 400 miles per working day.
How far does he drive in a year if he works on 235 days?

10 138 people attended a dinner dance.
The tickets were £26 each.
What was the total amount they paid?

Exercise 1.4H

1 Work out these without your calculator.
(a) 1^2 (b) 8^2 (c) 2^4
(d) 40^2 (e) 70^2 (f) 80^2
(g) 300^2 (h) 500^2 (i) 600^2
(j) 110^2

2 Use your calculator to work out these.
(a) 18^2 (b) 34^2 (c) 52^2
(d) 78^2 (e) 86^2 (f) 47^2
(g) 57^2 (h) 240^2 (i) 389^2
(j) 643^2

3 Use your calculator to work out these.
(a) 7^3 (b) 8^3 (c) 12^3
(d) 18^3 (e) 34^3 (f) 48^3
(g) 56^3 (h) 231^3

4 Use your calculator to work out these.
(a) $\sqrt{256}$ (b) $\sqrt{441}$ (c) $\sqrt{576}$
(d) $\sqrt{1024}$ (e) $\sqrt{3969}$ (f) $\sqrt{5476}$
(g) $\sqrt{3364}$ (h) $\sqrt{7921}$

5 Work out these without your calculator.
(a) $\sqrt{1600}$ (b) $\sqrt{4900}$ (c) $\sqrt{8100}$
(d) $\sqrt{10\,000}$ (e) $\sqrt{90\,000}$

Exercise 1.5H

1 Work out these.
(a) −3 add 5
(b) −8 add 4
(c) 7 subtract 11

2 The temperature is −7°C.
Find the new temperature after
(a) a rise of 4°C.
(b) a rise of 10°C.
(c) a fall of 6°C.

3 Find the difference in temperature between
(a) 8°C and 26°C.
(b) −3°C and 12°C.
(c) −23°C and −11°C.

4 Arrange these numbers in order, smallest first.
(a) 2, −6, 5, −4
(b) 7, −5, 0, −8
(c) −2, 6, −7, −8, 9, 3

 5 An aircraft flies at 15 000 m where the outside temperature is −70°C.
When the aircraft lands the outside temperature is 21°C.
What is the difference between these temperatures?

6 Work out these.
(a) 3 − 5 (b) 5 − 2 (c) 7 − 5
(d) 4 − 6 (e) 3 − 4 (f) 4 + 3
(g) −1 + 3 (h) −3 + 3 (i) −4 + 1
(j) 2 + 1 (k) −2 − 3 (l) −7 + 6
(m) −3 + 2 (n) −6 − 5 (o) −3 − 2
(p) −3 + (−4) (q) 4 + (−5) (r) −4 − (−3)
(s) −2 + (−1) (t) −6 − (−4)

Chapter 2

Algebra 1

Exercise 2.1H

1 (a) There are 6 men and 4 women on a bus.
   How many people are there on the bus?
   (b) There are 6 men and w women on a bus.
   Write an expression for the number of people on the bus.
   (c) There are m men and w women on a bus.
   Write an expression for the number of people on the bus.

2 (a) Shamsah buys 4 apples and 3 oranges.
   How many pieces of fruit does she buy?
   (b) Shamsah buys x apples and 3 oranges.
   Write an expression for the number of pieces of fruit she buys altogether.
   (c) Shamsah buys x apples and y oranges.
   Write an expression for the number of pieces of fruit she buys altogether.

3 (a) What is the length of this line?

   4 ———————— 7 - - - - - - - -

   (b) Write an expression for the length of this line.

   x ———————— 7 - - - - - - - -

   (c) Write an expression for the length of this line.

   4 ———————— y - - - - - - - -

4 (a) What is the length of this line?

   3 ———————— 5 - - - - - -

   (b) Write an expression for the length of this line.

   p ———————— 5 - - - - - -

   (c) Write an expression for the length of this line.

   3 ———————— q - - - - - -

5 (a) Write an expression for the length of this line.

   x ———————— x - - - - -

   (b) Write an expression for the length of this line.

   2 ———— 4 - - - - y ————

   (c) Write an expression for the length of this line.

   x ———— 4 - - - - y ————

6 (a) Write an expression for the length of this line.

   x ———— y - - - - 2 ————

   (b) Write an expression for the length of this line.

   x ———— 5 - - - - x ————

   (c) Write an expression for the length of this line.

   x ———— y - - - - x ————

7 The length of a rectangle is 6 cm longer than its width.
   (a) What is the length of the rectangle if the width is
   (i) 12 cm?   (ii) 16 cm?
   (b) Write an expression for the length of the rectangle if the width is w cm.

8 There are x people on a bus.
   At a bus stop, 5 more get on.
   Write an expression for the number of people now on the bus.

9 Cox apples cost 10p more a kilogram than Braeburn apples.
   (a) What is the cost of a kilogram of Cox apples when a kilogram of Braeburn apples costs
   (i) 85 pence?   (ii) 73 pence?
   (b) Write an expression for the cost of a kilogram of Cox apples when a kilogram of Braeburn apples costs x pence.

10 At Pete's chip shop, chips cost c pence a bag.
   Write an expression for the cost of
   (a) 2 bags.   (b) 4 bags.   (c) 8 bags.

Exercise 2.2H

1 Pam has 7 fewer books than James.
  (a) How many books does Pam have if James has
    (i) 12 books?   **(ii)** 20 books?
  (b) Write an expression for the number of books Pam has if James has *b* books.

2 Tea costs 20p a cup less than coffee.
  (a) What is the cost of a cup of tea if a cup of coffee costs
    (i) 70p?   **(ii)** 95p?
  (b) Write an expression for the cost of a cup of tea if a cup of coffee costs *c* pence.

3 Write an expression for the length of the dashed part of these lines.
  (a)
```
- - - - - - -       a
←————— 10 —————→
```
  (b)
```
- - - - - -         4
←——— b ———→
```
  (c)
```
       c
════════ - - - - - - - - - - - -
←——————— d ———————→
```

4 The width of a rectangle is 6 cm less than the length.
  (a) What is the width of the rectangle if the length is
    (i) 12 cm?   **(ii)** 16 cm?
  (b) Write an expression for the width of the rectangle if the length is *y* cm.

5 Mike earns £50 less per week than Adrian.
  (a) How much does Mike earn per week if Adrian earns
    (i) £160?   **(ii)** £340?
  (b) Write an expression for how much Mike earns per week if Adrian earns £*P*.

6 There were 24 people on a bus when it arrived at a stop. Some got off.
  (a) How many people were left on the bus if the number getting off was
    (i) 2?   **(ii)** 5?
  (b) Write an expression for the number of people left on the bus if the number getting off was *s*.

7 Josh bought a coat that cost £*x* less than the coat Jeremy bought.
  Write an expression, in pounds, for the cost of Josh's coat if Jeremy's cost
  (a) £65.   **(b)** £80.   **(c)** £*y*.

8 The width of a rectangle is half its length.
  (a) What is the width of the rectangle if the length is
    (i) 12 cm?   **(ii)** 16 cm?
  (b) Write an expression for the width of the rectangle if the length is *y* cm.

9 Write an expression for the length of the dashed part of these lines.

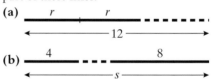

10 Joe spends £*x* in *d* days.
  He spends the same amount each day.
  Write an expression, in pounds, for the amount he spends in one day.

Exercise 2.3H

Simplify these expressions.

1 $a + a + a$

2 $b + b + b + b + b + b + b + b$

3 $c \times 6$

4 $2 \times d$

5 $5 \times c + 2 \times c$

6 $c + c + c + c - c + c$

7 $2a + 3a - 2a$

8 $5 \times b$

9 $3c - 2c$

10 $4a + 3a$

11 $6x + 4x - 2x$

12 $7a - a$

13 $4y - 3y$

14 $2b + 4b - b$

15 $2p - 3p + 4p$

16 $6s + 2s - 3s - 4s$

17 $b + 2b$

18 $a + 2a - 3a$

19 $3 \times b + 4 \times b$

20 $5a + 2a$

Exercise 2.4H

Simplify these expressions where possible.

1 $4a + 2b + a + b$

2 $3a + 2b - 2a$

3 $5 \times c \times d$

4 $5c + 3d$

5 $a + 2b + 4a + b$

6 $2 \times p + 3 \times q$

7 $3 \times a \times 4 \times b$

8 $2x + 4y + y + 2x$

9 $5 \times a \times a$

10 $6a + b + 3c + b + 3a + 3c$

11 $4a + 2a - 3a + a$

12 $2 \times a \times b + 4a \times a$

13 $3a - 2b$

14 $b \times b$

15 $4s + 2 - 3s + 3$

16 $a + 2b - a + 2b$

17 $2 \times a \times b \times b$

18 $3x + 2y + 4 + 2y - 2x + 3$

19 $a + 2b + 2a - 3b$

20 $2a \times 2b \times 3a \times b$

Exercise 3.1H

1 Which of the following are discrete data and which are continuous data?
> Time of day
> Weight of an egg
> Number of peas in a pod
> Day you were born
> Amount of rain in a week
> Price of an item

2 List at least three more examples of
   (a) continuous data.   **(b)** discrete data.

3 Design a data collection sheet for each of the following.
   (a) Favourite school subject   **(b)** Favourite sport   **(c)** School shoe colour

Exercise 3.2H

1 Draw a vertical line graph to show each of these sets of data.

(a)

Type of fast food	Frequency
Pizza	18
Chinese	8
Burger	13
Kebab	6
Fish and chips	4
Other	25

(b)

Type of vehicle	Frequency
Car	43
Van	29
Lorry	15
Bus	7
Motorbike	4
Other	2

(c)

Type of tree	Frequency
Oak	3
Beech	7
Pine	10
Sycamore	6
Lime	2
Other	2

2 Draw a bar chart to show each of these sets of data.

(a)

Favourite colour	Frequency
Blue	23
Red	9
Green	6
Yellow	4
Pink	7
Black	3
Other	8

(b)

Number of sisters	Frequency
0	7
1	12
2	8
3	2
4 or more	1

(c)

Shoe size	Frequency
39	12
40	8
41	15
42	11
43	9
44	6
45	2
Other	7

(d)

Type of bird	Frequency
Sparrow	41
Starling	32
Pigeon	14
Finch	5
Dove	2
Other	6

(e)

Drinks per day	Frequency
Less than 4	5
4	12
5	21
6	28
7	23
8 or more	11

(f)

Number of pets	Frequency
0	14
1	45
2	23
3	12
4	5
5 or more	1

Exercise 3.3H

 1 Here is a two-way table showing the results of a car survey.
Copy and complete the table.

	Black	Not black	Total
German	18	49	
Not German	23	64	
Total			

 2 Here is a two-way table showing the results of a survey of vehicles in a car park.
(a) Copy and complete the table.

	White	Not white	Total
Car	3	86	
Not Car	24	7	
Total			

(b) How many white cars were in the survey?
(c) How many cars were in the survey?
(d) How many vehicles were not white?
(e) How many vehicles were surveyed in total?

 3 Here is a two-way table showing the sales made by a bookshop on one day.
(a) Copy and complete the table.

	Paperback	Not paperback	Total
Fiction		11	98
Non-fiction	9	17	
Total			

(b) How many non-fiction books were sold?
(c) How many paperback fiction books were sold?
(d) How many books were sold in total?

4 A group of students were questioned about their favourite subject and their preferred writing hand.
Here is a two-way table showing the results.
(a) Copy and complete the table.

	Right hand	Left hand	Neither	Total
Maths	27	8	1	
Not maths		13		112
Total	126			

(b) How many students were left-handed?
(c) How many students chose maths?
(d) How many right-handed students chose a subject other than maths?
(e) How many students were in the survey?

5 At a swimming championship, Australia, China and the USA won most medals.
Their results are shown in the table.
(a) Copy and complete the table.

	Gold	Silver	Bronze	Total
Australia		27	13	74
China		20		63
USA	21	19		59
Total	71			

(b) How many silver medals were awarded in total to these countries?
(c) How many gold medals did Australia win?
(d) Which of these countries won most bronze medals?
(e) How many medals were awarded in total to these countries?

Exercise 3.4H

1 Draw a frequency table using tallies for each of these sets of data.

(a) Marks out of 30 in a maths test
   Use groups of 1 to 5, 6 to 10, 11 to 15, 16 to 20, ...

14	26	16	24	13	4	8	22	20	24
2	25	15	16	22	11	10	23	8	23
23	17	5	7	17	9	12	18	17	18
7	29	8	18	14	16	13	16	6	6
13	16	14	9	6	19	7	13	15	4
19	21	19	14	9	3	7	23	19	21

(b) Number of apples on a tree
   Use groups of 1 to 10, 11 to 20, 21 to 30, ...

43	34	17	37	18	42	11	37	18	15
24	25	23	23	30	24	13	16	28	21
15	37	35	27	28	33	21	27	6	18
23	2	29	39	31	40	36	86	27	46
7	25	21	19	90	29	27	24	14	27
36	21	30	25	22	31	29	37	7	43

(c) Number of lorries delivering to a depot per day
   Use groups of 0 to 19, 20 to 39, 40 to 59, ...

49	41	63	36	29	53	65	46	27	48
51	74	52	44	47	77	87	35	69	74
53	27	44	50	67	9	36	14	31	42
73	55	35	53	83	45	62	53	43	61
4	62	28	37	16	43	45	32	55	43
66	37	17	18	33	38	57	51	77	66

2 Draw a bar chart for each of these sets of data.

(a)

Number of video games	Frequency
1 to 5	7
6 to 10	14
11 to 15	31
16 to 20	27
21 to 25	11
26 or more	10

(b)

Number of swimmers	Frequency
0 to 49	2
50 to 99	5
100 to 149	11
150 to 199	9
200 to 249	3
250 or more	1

(c)

Number of eggs	Frequency
1 to 5	5
6 to 10	8
11 to 15	17
16 to 20	23
21 to 25	9
26 or more	3

(d)

Number of runners	Frequency
1 to 5	9
6 to 10	17
11 to 15	18
16 to 20	4
21 or more	2

(e)

Number of balloons	Frequency
0 to 9	4
10 to 19	11
20 to 29	27
30 to 39	35
40 to 49	21
50 to 59	13
60 to 69	5
70 or more	9

(f)

Number of fish	Frequency
1 to 10	9
11 to 20	9
21 to 30	15
31 to 40	28
41 to 50	7
51 or more	2

Exercise 4.1H

1 Write in words the place value of the digit 8 in each of these numbers.
(a) 800   (b) 0.8   (c) 8000
(d) 8.74   (e) 0.028

2 Write these numbers as decimals.
(a) $\frac{9}{10}$ (b) $8\frac{3}{10}$ (c) $\frac{7}{100}$
(d) $61\frac{23}{100}$ (e) $\frac{183}{1000}$

3 Write these decimals as fractions or mixed numbers (whole numbers and fractions) in their lowest terms.
(a) 0.4   (b) 5.1   (c) 27.5
(d) 0.25   (e) 1.13

4 Use the fact that there are 1000 millilitres in a litre to write these capacities in litres.
(a) 250 ml   (b) 1500 ml   (c) 185 ml
(d) 5 ml   (e) 2250 ml

5 Write these lengths in centimetres.
(a) 1.2 m   (b) 84 mm   (c) 3.85 m
(d) 56 mm   (e) 7 mm

6 Draw a grid to show that $\frac{2}{5} = \frac{4}{10} = 0.4$.

7 Write these fractions as decimals.
(a) $\frac{7}{10}$ (b) $\frac{1}{2}$ (c) $\frac{4}{5}$
(d) $\frac{3}{4}$ (e) $\frac{13}{100}$

Exercise 4.2H

1 Multiply each of these numbers by 10.
(a) 37   (b) 78   (c) 0.7
(d) 0.5   (e) 0.03

2 Multiply each of these numbers by 100.
(a) 37   (b) 78   (c) 0.7
(d) 0.5   (e) 0.03

3 Multiply each of these numbers by 1000.
(a) 37   (b) 78   (c) 0.7
(d) 0.5   (e) 0.03

4 Divide each of these numbers by 10.
(a) 73   (b) 21   (c) 67
(d) 29.7   (e) 5.26

5 Divide each of these numbers by 100.
(a) 73   (b) 21   (c) 67
(d) 29.7   (e) 5.26

6 Divide each of these numbers by 1000.
(a) 73   (b) 21   (c) 67
(d) 29.7   (e) 5.26

7 Work out these.
(a) $3.6 \times 1\,000\,000$ (b) $18.7 \times 10\,000$
(c) $821 \div 10\,000$ (d) $375.1 \div 100\,000$

8 On a model building, all distances are multiplied by 100 to find the distance on the real building.
The height of the model is 12 cm.
What is the height of the real building?

 9 On a road map, all real distances are divided by 100 000 to get the distance on the map.
What is the distance between two towns on the map that are 27 km apart in real life?

10 A microscope enlarges objects so that they appear to be 1000 times their real length.
An object is really 0.07 mm across.
How far across will it appear to be under the microscope?

Exercise 4.3H

1 Work out these.

(a)   3 . 8 7
     + 9 . 1 5
     ─────────

(b)   3 8 . 4 3
     + 5 9 . 1 2
     ─────────

(c)   4 1 . 5 3
     + 6 7 . 4 2
     ─────────

(d)   1 6 . 1 9
     +   8 . 3 4
     ─────────

(e)   3 6 . 8 4
     + 5 3 . 9 2
     ─────────

(f)   1 3 7 . 5 8
     + 2 7 3 . 1 6
     ─────────

2 Work out these.

(a)   1 9 . 2 8
     −   6 . 2 5
     ─────────

(b)   4 7 . 1 6
     − 1 5 . 4 2
     ─────────

(c)   2 5 3 . 8 0
     −   8 1 . 4 7
     ─────────

(d)   1 7 . 8 4
     −   8 . 1 3
     ─────────

(e)   5 7 . 4 2
     − 2 3 . 5 1
     ─────────

(f)   7 2 1 . 5 8
     − 2 3 6 . 8 2
     ─────────

3 Work out these.
(a) £3.95 + 82p + £1.57
(b) £15.21 + 77p + £3.42 + 61p
(c) £63.84 + 90p + £8.51 + £91.20 + 47p
(d) £6.25 + 42p + 87p + £63.20

4 Find the cost of six DVDs at £13.45 each.

5 In the javelin, Amina throws 52.15 m and Raj throws 46.47 m.
Find the difference between the lengths of their throws.

6 The times for the first and last places in a show-jumping event were 47.82 seconds and 53.19 seconds.
Find the difference between these times.

7 Melissa buys three of these bags of carrots.

 WEIGHT | PRICE
 0.450 KG | 99p

(a) What is the total weight?
(b) What is the total cost?

8 Find the cost of 3 kg of onions at £0.72 per kilogram.

9 Work out these.
Give your answers in the larger unit.
(a) 7.3 m + 87 cm + 6.1 m
(b) 5.2 m + 54 cm + 3.85 m + 76 cm
(c) 6.1 m − 278 cm
(d) 4.5 m − 57 cm
(e) 12.1 cm − 7 mm

10 Work out these.
Where applicable, give your answers in the larger unit.
(a) 650 g + 1.7 kg + 56 g + 2.1 kg
(b) 3.7 kg + 450 g − 1.5 kg
(c) 1.4 kg − 586 g
(d) 2 litres − 730 ml
(e) 3 × 0.58 litres

11 Holly buys two scarves at £8.45 each and a pair of shoes at £35.99.
How much change does she get from £60?

12 Dean buys two newspapers at 55p each and three magazines at £1.20 each.
How much change does he get from £10?

13 Work out these.
(a) 6×0.4
(b) 0.2×7
(c) 5×0.3
(d) 0.8×9
(e) 0.4×0.1
(f) 0.7×0.8
(g) 50×0.7
(h) 0.3×80
(i) 0.4×0.3
(j) 0.6×0.1
(k) $(0.7)^2$
(l) $(0.2)^2$

Chapter 5

Formulae

Exercise 5.1H

1 The total number of seats in an assembly hall is found by multiplying the number of rows by 20. How many seats are there when there are
   (a) 20 rows?   (b) 15 rows?
   (c) 17 rows?   (d) $13\frac{1}{2}$ rows?

2 The perimeter of a square is found by multiplying the length of one side by 4.
   Work out the perimeter of a square with sides of these lengths.
   (a) 5 cm   (b) 13 cm
   (c) $6\frac{1}{2}$ cm   (d) 8.2 cm

3 The cost of a child's bus fare is half the cost of an adult's fare.
   What is the cost of a child's fare when the adult fare is
   (a) 80p?   (b) 70p?
   (c) £1.30?   (d) £2.36?

4 Five friends divide a bag of sweets equally.
   To work out how many sweets each person receives, divide the total number of sweets by 5.
   How many sweets does each person receive when the bag contains
   (a) 20 sweets?   (b) 55 sweets?
   (c) 80 sweets?   (d) 125 sweets?

5 The time, in minutes, needed to cook a piece of beef can be found by multiplying the weight of the beef, in kilograms, by 40.
   How long does it take to cook a piece of beef weighing
   (a) 2 kg?   (b) 8 kg?
   (c) $5\frac{1}{2}$ kg?   (d) 3.2 kg?

6 The area of a rectangular room is found by multiplying the length by the width.
   Work out the area of these rooms.
   (a) Length 4 m and width 3 m
   (b) Length and width both 7 m
   (c) Length $2\frac{1}{2}$ m and width 6 m
   (d) Length 5.4 m and width 5 m

7 A car hire firm charges a fee of £30, plus £2 for each mile travelled.
   How much is the total charge if you travel
   (a) 25 miles?   (b) 70 miles?
   (c) 400 miles?   (d) 185 miles?

8 A football team earns 3 points for every match they win and 1 point for every match they draw.
   How many points in total do they earn when they
   (a) win 3 games and draw 5?
   (b) win 7 games and draw 10?
   (c) win 14 games and draw 6?
   (d) win 28 games and draw 3?

9 To change American dollars into pounds, divide the number of dollars by 1.5.
   How many pounds will you get for
   (a) $6?   (b) $150?
   (c) $24?   (d) $10.50?

10 To change a distance from miles into kilometres, multiply the number of miles by 8 and divide by 5.
   Work out the number of kilometres that is the same as
   (a) 10 miles.   (b) 15 miles.
   (c) 100 miles.   (d) 44 miles.

Exercise 5.2H

In questions **1** to **7**, write down a formula for each situation using the letters in **bold**.

1 The total number of **s**eats in an assembly hall is found by multiplying the number of **r**ows by 20.

2 The **p**erimeter of a square is found by multiplying the **l**ength of one side by 4.

3 The cost of a **c**hild's bus fare is half the cost of an **a**dult's fare.

4 Five friends divide a bag of sweets equally. To work out how many sweets **e**ach person receives, divide the total number of **s**weets by 5.

5 The **t**ime, in minutes, needed to cook a piece of beef can be found by multiplying the **w**eight of the beef, in kilograms, by 40.

6 The **a**rea of a rectangular room is found by multiplying the **l**ength by the **w**idth.

7 A car hire firm **c**harges a fee of £30, plus £2 for each **m**ile travelled.

In questions **8** to **10**, write down a formula for the situation using appropriate letters and say what each letter stands for.

8 A football team earns 3 points for every match they win and 1 point for every match they draw.

9 To change American dollars into pounds, divide the number of dollars by 1.5.

10 To change a distance from miles into kilometres, multiply the number of miles by 8 and divide by 5.

Exercise 5.3H

1 Find the value of these expressions when $a = 4$ and $b = 2$.
   (a) $a + b$ (b) $a - b$ (c) $2a$
   (d) $7a$ (e) $5b$ (f) $8b$
   (g) ab (h) a^2 (i) $3ab$
   (j) $2a + b$ (k) $3a - b$ (l) $5a + 4b$
   (m) $3a - 4b$ (n) $a^2 + b^2$ (o) $3a^2$
   (p) $6b^2$ (q) a^2b (r) b^3
   (s) $\dfrac{a}{b}$ (t) $\dfrac{10b}{a}$ (u) $b - a$
   (v) $\dfrac{4b}{2a}$ (w) $\dfrac{a^2}{b^2}$ (x) $\dfrac{b}{a}$

2 Use the formula $C = 5t + 2$ to find C when
   (a) $t = 2$.   (b) $t = 8$.   (c) $t = 12$.
   (d) $t = 20$.   (e) $t = \frac{1}{2}$.

3 Use the formula $P = 2C - F$ to find P when
   (a) $C = 4$, $F = 4$.   (b) $C = 3$, $F = 5$.
   (c) $C = 8$, $F = 5$.   (d) $C = 0$, $F = 7$.

4 Use the formula $X = w + nd$ to find X when
   (a) $w = 1$, $n = 2$, $d = 1$.
   (b) $w = 4$, $n = 2$, $d = 6$.
   (c) $w = 12$, $n = 10$, $d = 0$.
   (d) $w = 50$, $n = 30$, $d = 10$.
   (e) $w = 11$, $n = 9$, $d = \frac{1}{2}$.

5 Find the value of these expressions when $k = 3$ and $m = 2$.
   (a) $k + m$ (b) $5k$ (c) km
   (d) $m - k$ (e) $4k + 6m$

6

Equations 1

Exercise 6.1H

Solve these equations.

1 $x - 5 = 12$

2 $5x = 40$

3 $9y = 45$

4 $x + 7 = 18$

5 $10p = 60$

6 $2x = 2$

7 $3x = 18$

8 $6y = 48$

9 $x - 2 = -5$

10 $x - 12 = -4$

11 $x - 10 = -2$

12 $x - 11 = -11$

Exercise 6.2H

Solve these equations.

1 $15 - x = 9$

2 $7 - x = 10$

3 $3x + 2 = 11$

4 $4x + 1 = 17$

5 $3x - 5 = 7$

6 $3x + 6 = 48$

7 $3x - 7 = 41$

8 $8x - 3 = 29$

9 $4x + 12 = 20$

10 $4x + 18 = 22$

11 $5x + 6 = 31$

12 $2x - 4 = -2$

Exercise 6.3H

Solve these equations.

1 $\frac{x}{4} = 7$

2 $\frac{a}{3} = 6$

3 $\frac{p}{7} = 7$

4 $\frac{y}{4} = 12$

5 $\frac{x}{2} = 18$

6 $\frac{x}{4} = 6$

7 $\frac{x}{3} = 7$

8 $\frac{x}{5} = 5$

9 $\frac{x}{6} = 1$

10 $\frac{x}{3} = 3$

11 $\frac{x}{3} = 9$

12 $\frac{x}{4} = 20$

Exercise 6.4H

1 Robbie thinks of a number.
   He adds 3 to it. The answer is 8.
   Use x to represent Robbie's number.
   Write down an equation and solve it to find
   Robbie's number.

2 Patsy thinks of a number.
   She multiplies it by 3. The answer is 15.
   Use x to represent Patsy's number.
   Write down an equation and solve it to find
   Patsy's number.

3 Four chocolate biscuits cost 64p.
   Use c to represent the cost of one biscuit in
   pence.
   Write down and solve an equation to find the
   cost of one biscuit.

4 Patel has 16 CDs. Sonia has x CDs.
   Together they have 34 CDs.
   Write down and solve an equation to find how
   many CDs Sonia has.

5 Joy had x pens.
   She gives away 6 of them and has 7 left.
   Write down and solve an equation to find how
   many pens Joy had to start with.

6 Francis buys three packets of mints and pays with
   a pound coin.
   He gets 16p change.
   Use m to represent the cost of a packet of mints
   in pence.
   Write down and solve an equation to find the
   cost of one packet of mints.

Coordinates

Exercise 7.1H

1 Write down the coordinates of the points A, B, C, D, E, F, G, H, I and J.

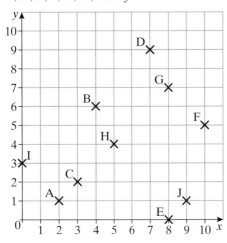

2 On a grid, draw x- and y-axes from 0 to 8. Plot and label these points.

A(7, 1)   B(3, 5)   C(4, 2)   D(2, 0)   E(0, 6)

Exercise 7.2H

1 Write down the coordinates of the points A, B, C, D, E, F, G, H, I and J.

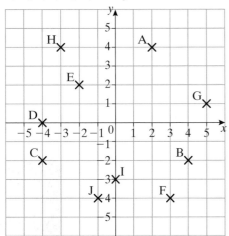

In questions **2** to **4** you will need to draw x- and y-axes from −5 to 5.

2 Plot and label the points A(5, 1), B(5, −3), C(−3, −3) and D(−3, 1).
Join the points to make the shape ABCD.

3 Plot and label the points A(4, 1), B(4, −4) and C(−3, −4).
Join the points to make the shape ABC.

4 Plot and label the points A(2, 4), B(2, −2), C(−1, −4) and D(−1, 2).
Join the points to make the shape ABCD.

Exercise 7.3H

1 Write down the equation of each of the lines **(a)**, **(b)**, **(c)** and **(d)**.

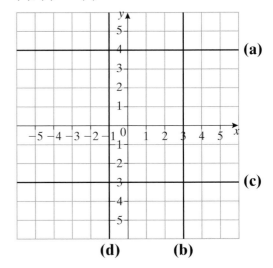

Statistical calculations 1

Exercise 8.1H

1 Find the median of each of these sets of data.
  (a) 1   4   5   7   9   9   10
  (b) 3   9   6   4   3   9   3   2   7
  (c) 4   5   9   6   9   9   7   4   8   8
  (d) 2   5   3   2   7   8   9   3   4
  (e) 3   7   1   3   5   7   9   11

2 Find the median of these seven weekly wages.
  £257   £238   £289   £362
  £321   £411   £306

3 The ages of six teachers are as follows.
  46   52   38   25   31   62
  Find the median age.

4 These are Sophie's marks in five tests.
  5   8   9   7   8
  These are Carole's marks in six tests.
  7   6   9   4   8   3
  (a) Find the median mark for
      (i) Sophie.   **(ii)** Carole.
  (b) Who had the higher median mark?

5 Stephen is asked to find the median of these numbers.
  4   2   1   5   7
  He says the median is 1. He is wrong.
  (a) What has he done wrong?
  (b) What is the correct answer?

Exercise 8.2H

1 Find the mode of this set of data.
  1   2   1   3   2   4   3   5   2   5   2

2 The number of boys who were at ten different sports practices were as follows.
  18   17   17   20   16
  20   9   9   17   10
  Find the modal number of boys at a practice.

3 The marks scored in a test were as follows.
  10   12   18   17   16   18   14   15
  16   18   14   18   20   9   18   13
  Find the modal number of marks scored.

4 Here is a list of the heights of ten people.
  173 cm   158 cm   161 cm   163 cm
  181 cm   153 cm   173 cm   170 cm
  162 cm   180 cm
  Find the modal height.

5 The ages of a group of friends are as follows.
  18   22   21   19   18
  18   17   19   17   21
  Find the modal age.

Exercise 8.3H

1 Find the mean of this set of data.

    3   6   5   4   7   6   8   1

2 A gardener measures the heights of a group of plants.
The heights were as follows.

    50 cm   65 cm   80 cm   40 cm   35 cm

Find the mean height.

3 In a test the marks were as follows.

    9   7   8   7   5   6   8   9   5   6

Find the mean mark.

4 The ages of a group of friends are as follows.

    18   22   21   19   18
    18   17   19   17   21

Find the mean age.

5 The salaries of ten workers in a small company are as follows.

£20 000	£20 000
£20 000	£18 000
£22 000	£23 000
£25 000	£21 000
£23 000	£28 000

Find the mean salary.

Exercise 8.4H

1 Find the range of this set of data.

    3   6   5   4   7   6   8   1

2 A gardener measures the heights of a group of plants.
The heights were as follows.

    50 cm   65 cm   80 cm   40 cm   35 cm

Find the range of the heights.

3 In a test the marks were as follows.

    9   7   8   7   5   6   8   9   5   6

Find the range of the marks.

4 The ages of a group of friends are as follows.

    18   22   21   19   18
    18   17   19   17   21

Find the range of their ages.

5 The salaries of ten workers in a small company are as follows.

£20 000	£20 000
£20 000	£18 000
£22 000	£23 000
£25 000	£21 000
£23 000	£28 000

Find the range of their salaries.

Exercise 8.5H

1 The table shows the weights of the women in a keep fit group.

Weight in kg (to the nearest kilogram)	Number of women
45–49	8
50–54	12
55–59	10
60–64	6
65–69	8

Write down the modal class.

2 The table shows the weekly pay of 30 workers.

Weekly pay (£)	Number of workers
100–149.99	3
150–199.99	10
200–249.99	12
249–299.99	5

Write down the modal class.

3 The ages of the people at a bridge club are shown in this table.

Age	Frequency
20–29	5
30–39	12
40–49	8
50–59	15
60–69	26
70–79	10

Find the modal class.

4 The marks gained in a test are shown below.

55	60	62	74	53
59	73	81	91	48
43	62	90	85	45
63	67	75	49	84
61	67	44	68	77
83	49	84	76	63
53	73	88	64	74
79	56	63	44	59
82	59	65	57	87
83	72	70	51	48

Copy and complete the frequency table and write down the modal class.

Mark	Tally	Total
41–50		
51–60		
61–70		
71–80		
81–90		
91–100		

Exercise 9.1H

1 Find the first four terms of each of these
sequences.
(a) First term 3, term-to-term rule Add 5
(b) First term −5, term-to-term rule Add 2
(c) First term 23, term-to-term rule Subtract 5
(d) First term 60, term-to-term rule Subtract 25
(e) First term −17, term-to-term rule Add 6
(f) First term −1, term-to-term rule Add $\frac{1}{2}$

2 Write down the next two terms in each of these
sequences and give the term-to-term rule.
(a) 1, 8, 15, 22, 29, 36, ...
(b) 7, 13, 19, 25, 31, 37, ...
(c) 9, 17, 25, 33, 41, 49, ...
(d) 4, 13, 22, 31, 40, 49, ...
(e) 16, 21, 26, 31, 36, 41, ...
(f) 23, 30, 37, 44, 51, 58, ...

3 Write down the next two terms in each of these
sequences and give the term-to-term rule.
(a) 41, 35, 29, 23, 17, 11, ...
(b) 39, 32, 25, 18, 11, 4, ...
(c) 26, 21, 16, 11, 6, 1, ...
(d) 29, 23, 17, 11, 5, −1, ...
(e) 43, 35, 27, 19, 11, 3, ...
(f) 68, 55, 42, 29, 16, 3, ...

4 Find the missing numbers in each of these
sequences and give the term-to-term rule.
(a) 1, 7, ..., 19, 25, ...
(b) 11, ..., 35, 47, ..., 71
(c) 85, ..., 67, ..., 49, 40
(d) ..., 87, 83, 79, ..., 71
(e) 7, 22, ..., ..., 67, 82
(f) 34, ..., 20, 13, 6, ...

Exercise 9.2H

 1 Draw the next pattern in each of these sequences.
For each sequence, count the dots in each pattern
and find the term-to-term rule.

(a)

(b)

(c)

(d)

 2 Draw the next pattern in each of these sequences.
For each sequence, count the lines in each
pattern and find the term-to-term rule.

(a)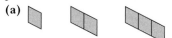

(b)

(c)

(d)

Exercise 9.3H

1 Write down the first five terms of the sequences
with these nth terms.
(a) $7n − 3$ (b) $9n − 6$ (c) $13n − 11$
(d) $2n + 61$ (e) $4n + 39$ (f) $3n + 13$

2 Find the 100th term of the sequences with these
nth terms.
(a) $4n + 4$ (b) $3n + 34$ (c) $13n − 8$
(d) $9n − 1$ (e) $5n + 38$ (f) $6n + 19$

Chapter 10

Measures

Exercise 10.1H

1 What are the readings on this scale?

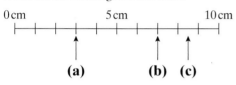

2 What is the reading on this scale?

3 What are the readings on this thermometer?

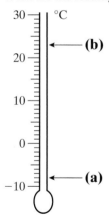

4 How long is this nail?

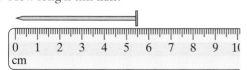

5 Measure the distance between A and B.

Exercise 10.2H

1 Change these lengths to centimetres.
   (a) 2 m **(b)** 3.5 m **(c)** 20 mm
   (d) 15 m **(e)** 45 mm

2 Change these lengths to millimetres.
   (a) 2 cm **(b)** 5.5 cm **(c)** 10 cm
   (d) 2 m **(e)** 3.5 m

3 Change these weights to grams.
   (a) 2 kg **(b)** 5 kg **(c)** 6.35 kg
   (d) 0.8 kg **(e)** 0.525 kg

4 Put these volumes in order, smallest first.
   1.2 litres   500 ml   2 litres   2500 ml   800 cl

5 Which metric units would you use to measure these lengths?
   (a) The span of your hand
   (b) The length of a corridor
   (c) The width of a window
   (d) The distance you can walk in a day
   (e) The distance from London to Edinburgh

6 Put these weights in order, smallest first.
   1.2 kg   1500 g   160 g   2000 g   0.8 kg

7 Change these volumes to millilitres.
   (a) 2 litres **(b)** 3.5 litres **(c)** 2 cl
   (d) 15 cl **(e)** 0.345 litres

8 Graham has three pieces of string.
   The lengths are 45 cm, 85 mm and 1.2 m.
   (a) Write them down in order, shortest first.
   (b) What is the total length of string
      (i) in millimetres?
      (ii) in centimetres?
      (iii) in metres?

Exercise 10.3H

Here are some approximate conversions between imperial and metric units.

Length	Weight
8 km ≈ 5 miles	1 kg ≈ 2 pounds (lb)
1 m ≈ 40 inches	
1 inch ≈ 2.5 cm	**Capacity**
1 foot (ft) ≈ 30 cm	4 litres ≈ 7 pints (pt)

1 Change these measures from imperial units to their approximate metric units.
   (a) 6 feet   (b) 35 miles   (c) 14 lb
   (d) 45 lb   (e) 14 pints

2 Change these measures from metric units to their approximate imperial units.
   (a) 24 km   (b) 5 m   (c) 5 kg
   (d) 12 litres   (e) 20 cm

3 Stephen needs 1 pound of meat for a recipe. How many grams should he buy?

4 A radiator is 20 inches wide. What is this in centimetres?

5 Cola is sold in 2 litre bottles. How much is this in pints?

6 Pauline is 5 foot 4 inches tall. What is this in centimetres?

7 A more accurate conversion between kilograms and pounds is 1 kg = 2.2 lbs. Sarfraz weighs 45 kg.
   (a) What is this in pounds?
   Many people still talk about how much they weigh in stones and pounds.
   There are 14 pounds in a stone.
   (b) What does Sarfraz weigh in stones and pounds?

8 Mark drives towards Nottingham and sees this sign.

Grantham	23 miles
Nottingham	53 miles

Roughly how far is it, in kilometres, from Grantham to Nottingham?

Exercise 10.4H

1 Estimate the following.
   (a) The height of your desk
   (b) The width of a door
   (c) The mass of a glass of water
   (d) The capacity of a kitchen sink
   (e) The distance you can walk in an hour

2 Estimate the height of this lamp post.

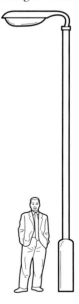

3 The building in this picture is about 16 m high.
Estimate the height of the lorry.

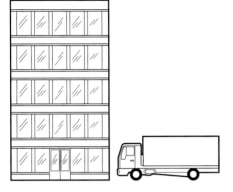

4 Alan wants to estimate the length of his
bedroom. His foot is 20 cm long.
He works out that the room is 16 foot lengths
long.
What is his estimate of the length of his bedroom?

5 Adrian is asked to estimate the distance he walks
to school.
His answer is 1.135 km.
(a) Why is this not a sensible estimate?
(b) What would be a better estimate?

11 Constructions 1

Chapter 11

Exercise 11.1H

1 Measure the length of each of these lines in centimetres.

(a) ————————————————————————

(b) ————————

(c) ——————————————————

(d) ——————————

(e) ————————————————

2 Measure the length of each of these objects.

(a)

(b)

3 Measure the length of each side of these shapes.

(a) 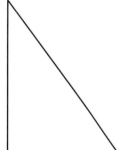   (b)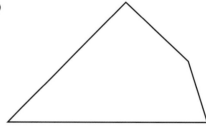

Exercise 11.2H

1 Match each of the following angles to its diagram.

295°   45°   168°   25°   90°

(a)   (b)   (c)   (d)   (e)

2 Are these angles acute, right angle, obtuse or reflex?

(a)

(b)

(c)

(d)

(e)

(f)

(g)

(h)

3 Are angles of these sizes acute, right angle, obtuse or reflex?
   (a) 57°   (b) 110°   (c) 90°   (d) 195°   (e) 28°
   (f) 124°   (g) 345°   (h) 91°   (i) 222°   (j) 6°

Exercise 11.3H

Copy and complete this table.
Estimate each angle first and then measure it with a protractor.

Angle	Estimated size	Measured size
a		
b		
c		
d		
e		
f		
g		
h		
i		
j		

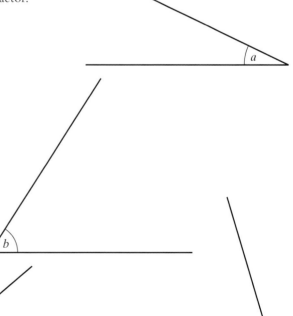

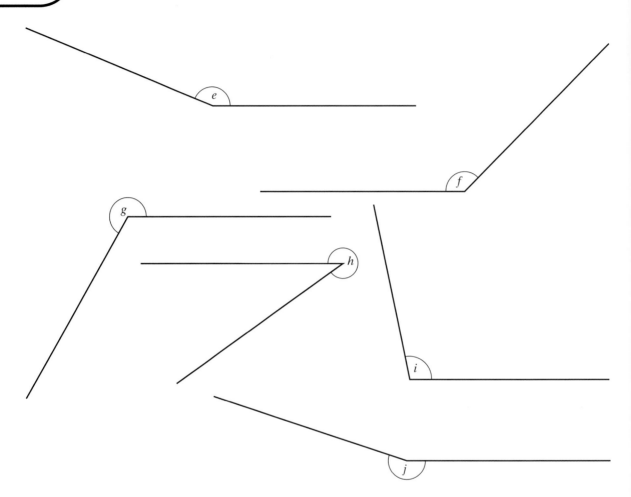

Exercise 11.4H

1 Draw accuractely each of these angles.

(a)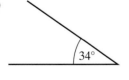

(b)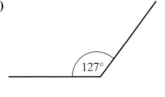

(c)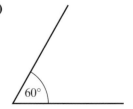

2 Draw accurately each of these angles.

(a) 70°	**(b)** 45°	**(c)** 151°	**(d)** 67°
(e) 139°	**(f)** 28°	**(g)** 80°	**(h)** 113°

3 Draw accurately each of these reflex angles.

(a) 210°	**(b)** 300°	**(c)** 188°	**(d)** 275°

Exercise 11.5H

1 Make an accurate full-size drawing of each of these triangles. For each triangle, measure the unknown length and angles from your drawing.

(a)

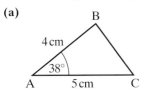

(b)

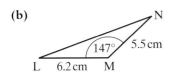

(c) Triangle ABC where AB = 8 cm, angle BAC = 58° and AC = 5 cm.

2 Make an accurate full-size drawing of each of these triangles. For each triangle, measure the unknown lengths and angle from your drawing.

(a)

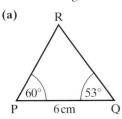

(b) C

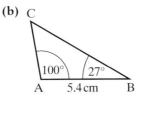

(c) Triangle XYZ where YZ = 6.5 cm, angle XZY = 67° and angle ZYX = 43°.

Exercise 11.6H

1 Make an accurate full-size drawing of each of these triangles. For each triangle, measure all the angles from your drawing.

(a)

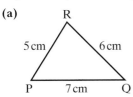

(b)

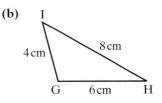

(c) Triangle ABC where AB = 6 cm, BC = 6.5 cm and AC = 7 cm.

2 Make an accurate full-size drawing of each of these triangles. For each triangle, measure the unknown length and angles from your drawing.

(a) N   **(b)**

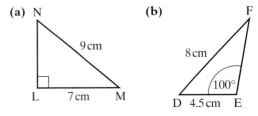

(c) Triangle XYZ where YZ = 6.8 cm, angle XZY = 42° and XY = 8 cm.

Exercise 11.7H

1 Measure each of these lines as accurately as possible.
Using the scales given, work out the length that each line represents.

(a) ————————————
    1 cm to 6 m

(b) ——————————
    1 cm to 20 km

(c) ————————————————
    2 cm to 5 miles

(d) ——————————
    1 cm to 4 m

2 Draw accurately the line to represent these actual lengths.
Use the scale given.
(a) 6 m   Scale: 1 cm to 1 m
(b) 8 km   Scale: 1 cm to 2 km
(c) 15 miles   Scale: 3 cm to 5 miles
(d) 450 m   Scale: 1 cm to 100 m

3 Here is a plan of a bedroom.
The scale of the drawing is 1 cm to 50 cm.

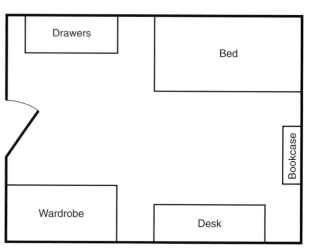

(a) Work out the length and width of the bedroom.
(b) Work out the length and width of each of the five items in the bedroom.
(c) The window in the bedroom measures 1 m by 1 m 75 cm.
What will the measurements of the window be to this scale?

4 The map shows some towns and cities in Scotland.
The scale of the map is 1 cm to 10 km.

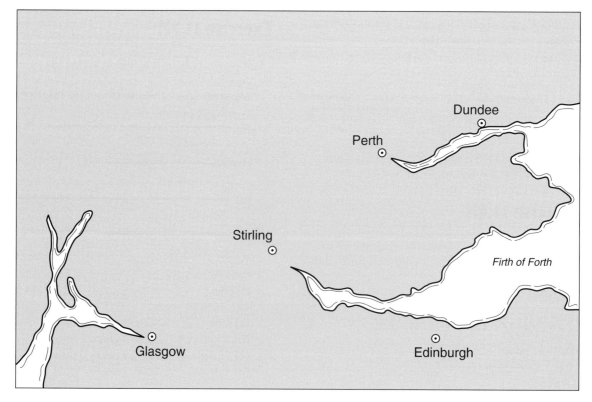

(a) What is the real–life distance, in kilometres, between these towns?
   (i)   Glasgow and Stirling
   (ii) Edinburgh and Glasgow
   (iii) Edinburgh and Perth
   (iv) Glasgow and Dundee
   (v)   Perth and Dundee
   (vi) Edinburgh and Stirling
(b) It is 660 km from Edinburgh to London.
   How many centimetres will this be to the same scale?

Exercise 11.8H

1 Measure the bearing of each of the classrooms in the map of a school from the playground.
   How many metres is each place from the playground?
   The scale of the map is 1 cm to 10 m.

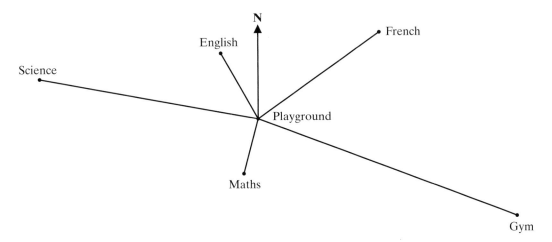

2 Three towns are Delham, Easton and Fanbury.
   Easton is 3 miles from Delham on a bearing of 065°.
   Fanbury is $2\frac{1}{2}$ miles from Delham on a bearing of 290°.
   Make a scale drawing showing these three towns.
   Use a scale of 2 cm to 1 mile.

3 Jenny went for a walk.
   She started from home (H) and walked 500 m on a bearing of 150° to the zoo (Z).
   From the zoo she walked 900 m on a bearing of 088° to the café (C).
   From the café she walked 600 m on a bearing of 350° to the shop (S).
   (a) Draw an accurate scale drawing of Jenny's walk.
       Use a scale of 1 cm to 100 m.
   (b) How far is Jenny from home?
   (c) On what bearing must she walk to get back home?

Chapter 12 — Using a calculator

Exercise 12.1H

1 Round these numbers to the nearest whole number.
(a) 24.3   (b) 561.5   (c) 81.2
(d) 417.6   (e) 81.91   (f) 128.43
(g) 39.8   (h) 627.203   (i) 5.32
(j) 4.57

2 Round these numbers to 1 decimal place.
(a) 2.37   (b) 8.24   (c) 9.45
(d) 12.6666   (e) 8.912   (f) 12.87
(g) 9.624   (h) 7.465   (i) 6.1919
(j) 1.95

3 Round these numbers to 2 decimal places.
(a) 2.837   (b) 1.624   (c) 5.465
(d) 5.1919   (e) 1.2345   (f) 7.381
(g) 6.247   (h) 7.318   (i) 7.904
(j) 9.595

4 Round these numbers to 1 significant figure.
(a) 64   (b) 372   (c) 9414
(d) 98   (e) 9012   (f) 54217
(g) 2594   (h) 45092   (i) 1631
(j) 1871

5 A man won £3871242 in a lottery.
Write this correct to 1 significant figure.

Exercise 12.2H

1 Work out these squares.
(a) 4.3^2 (b) 7.2^2 (c) 56^2
(d) 419^2 (e) 0.74^2 (f) 0.82^2
(g) 0.09^2 (h) 4.71^2 (i) 63.8^2

2 Work out these square roots.
(a) $\sqrt{14.44}$ (b) $\sqrt{68.89}$ (c) $\sqrt{3.61}$
(d) $\sqrt{219.04}$ (e) $\sqrt{3844}$ (f) $\sqrt{0.1156}$
(g) $\sqrt{0.5776}$ (h) $\sqrt{0.0324}$ (i) $\sqrt{10.3041}$

3 Work out these square roots.
Give your answers to 2 decimal places.
(a) $\sqrt{45.3}$ (b) $\sqrt{74.8}$ (c) $\sqrt{44}$
(d) $\sqrt{827}$ (e) $\sqrt{5632}$ (f) $\sqrt{0.468}$
(g) $\sqrt{0.4}$ (h) $\sqrt{56846}$ (i) $\sqrt{0.408}$
(j) $\sqrt{0.063}$

4 The area of a square is 720 cm².
Find the length of the side.
Give your answer to 1 decimal place

5 Work out these.
Give your answers to 2 decimal places.
(a) 2.4^4 (b) 1.83^6 (c) 3.59^5

Exercise 12.3H

Work out these on your calculator.
If the answers are not exact, give them correct to 3 decimal places.

1 $(7.3 + 3.2) \div 4.8$

2 $(134 - 43) \div 35$

3 $(8.2 - 3.6) \times 5.4$

4 $\sqrt{(17.3 + 16.8)}$

5 $\sqrt{(68.7 - 2.3^2)}$

6 $(7.3 - 2.6)^2$

7 $7.3^2 - 2.6^2$

8 $16.8 \div (5.2 - 1.9)$

9 $5.8 \times (1.9 + 7.3)$

10 $\sqrt{(28.6 - 9.7)}$

11 $\sqrt{(26.2 \div 3.8)}$

12 $34.9 \div (2.8 + 5.3)$

Chapter 13 Statistical diagrams 1

Exercise 13.1H

1 Draw a pie chart for each of these sets of data.

(a)

Favourite drink	Frequency
Soft drink	42
Milk	12
Water	48
Juice	18
Other	60
Total	180

(b)

Eye colour	Frequency
Blue	76
Blue/Green	28
Brown	92
Grey	38
Other	6
Total	240

(c)

Type of programme	Frequency
Comedy	14
Soap	22
Cartoon	16
Drama	18
Other	10
Total	80

2 Draw a pie chart for each of these sets of data.

(a)

Activity	Hours
School	6
Sleeping	9
Eating	2
Playing	3
TV	2
Other	2

(b)

Favourite TV channel	Frequency
BBC	12
ITV1	15
Channel 4	6
Five	9
Satellite/Cable	30

(c)

Favourite type of film	Frequency
Comedy	25
Horror	14
Romance	32
Action	17
Other	2

Exercise 13.2H

1 The pie chart shows the favourite flavour of crisp of 60 people questioned in a survey.

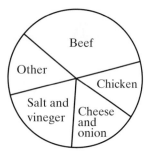

How many people preferred
(a) beef flavour?
(b) cheese and onion flavour?
(c) salt and vinegar flavour?

2 The pie chart shows the eye colour of 108 people questioned in a survey.

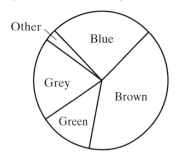

How many people had
(a) blue eyes?   **(b)** grey eyes?

3 The pie chart shows the distribution of the population of Britain.

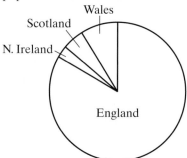

The total population of Britain is approximately 60 million people.
What is the approximate population of
(a) England?
(b) Wales?
(c) Northern Ireland?
(d) Scotland?

4 The pie chart shows the approximate land area of Britain.

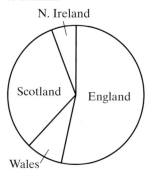

The total land area of Britain is approximately 94 320 square miles.
What is the approximate land area of
(a) England?
(b) Wales?
(c) Northern Ireland?
(d) Scotland?
Give your answers to 1 significant figures.

5 The pie chart shows the ingredients needed to make fruit crumble.
Betty makes a lage fruit crumble weighing 2.4 kg.
What is the approximate weight, in grams, of each of the ingredients
of Betty's crumble?

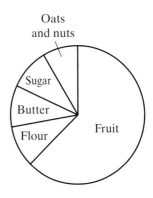

Exercise 13.3H

1 The table shows the maximum daytime temperature in Miami over a period of 12 days.

Day	Mon	Tues	Wed	Thur	Fri	Sat	Sun	Mon	Tues	Wed	Thur	Fri
Temp. (°C)	27	29	28	31	33	34	37	36	33	32	30	31

Draw a line graph to show this information.

2 The table shows the monthly sales of gravel by a garden centre in 2009.

Month	Jan	Feb	Mar	Apr	May	June	July	Aug	Sept	Oct	Nov	Dec
Sales (tonnes)	2	3	5	8	13	11	6	5	7	8	4	1

Draw a line graph to show this information.

3 The table shows the number of tickets sold by a cinema one week.

Day	Mon	Tues	Wed	Thur	Fri	Sat	Sun
Tickets	58	71	49	86	183	205	152

Draw a line graph to show this information.

4 The table shows the monthly sales of widgets in 2009.

Month	Jan	Feb	Mar	Apr	May	June	July	Aug	Sept	Oct	Nov	Dec
Sales (× £1000)	3	5	8	13	24	36	29	18	13	9	6	4

Draw a line graph to show this information.

Exercise 13.4H

1 The temperature of some water was taken every
5 minutes as it was heated.
   (a) What was the temperature of the water
   before it was heated?
   (b) How many minutes did it take for the
   liquid to reach 75°C?
   (c) How many degrees did the water
   temperature rise in the first 20 minutes?
   (d) What happened after 30 minutes?

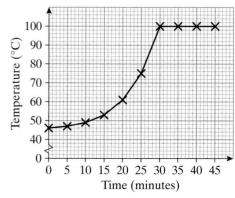

 2 The line graph shows the number of spectators
at the first six games played by Rangers and by
Rovers in the season.
   (a) How big was the crowd at Rovers'
   second game?
   (b) For which game did both teams have
   approximately the same size crowd?
   (c) What was the range of the crowd size for
   the Rangers games?
   (d) What was the mean crowd size for the
   Rovers games?
   (e) Can you use this information to predict
   crowd sizes for game 7? Explain your answer.

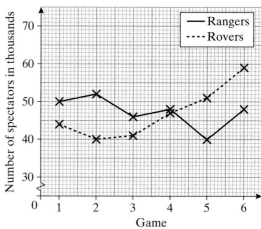

 3 The line graph below shows the total rainfall in
Bananaland theme park one year.
   (a) How much rain fell in June?
   (b) Which month had the most rain?
   (c) What was the range of the monthly rainfall
   figures for the year?
   (d) Calculate the mean monthly rainfall for the year.

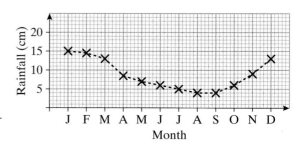

4 The line graph shows the value of sales of a
mushroom farm one year.
 (a) Which month had the lowest sales?
 (b) What was the range of sales values for the year?
 (c) What was the total value of sales for the year?
 (d) What was the mean value of sales for a month?

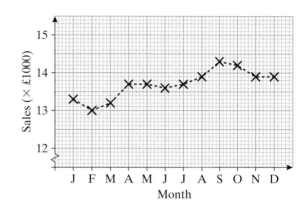

5 The line graph shows the number of
drivers paying to cross a toll bridge
one week.
 (a) How many drivers crossed the
 bridge during the week?
 (b) What was the mean number of
 drivers crossing the bridge each day?
 (c) What was the range of the number
 of drivers crossingthe bridge each day?
 (d) Why might there be fewer drivers
 crossing the bridge on Saturday
 and Sunday?

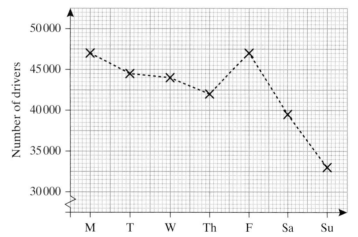

6 The line graph shows the mean
monthly rainfall and mean maximum
daily temperature in Brisbane one year.
 (a) Which months had the highest
 mean monthly rainfall?
 (b) Which months had the highest
 maximum daily temperature?
 (c) What was the maximum daily
 temperature in September?
 (d) In which month was the mean
 monthly rainfall approximately
 14 cm?
 (e) What was the range of the
 maximum daily temperature
 for the year?
 (f) What was the range of the mean
 monthly rainfall for the year?

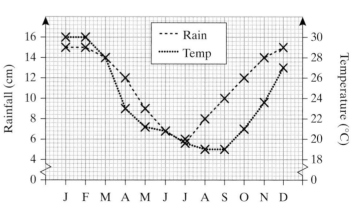

Exercise 13.5H

1 The pie charts show the number of companies involved in the automotive trade in the UK in 2006 and 2009. The total number of companies in each year was the same.

Number of companies in 2006

Number of companies in 2009

    (a) In which year was the number of car repair companies larger?
    (b) What is the main difference between the data for the two years?
    (c) What similarities are there between the data for the two years?

2 The diagrams below show the performances of boys and girls in a general knowledge test.

Girls

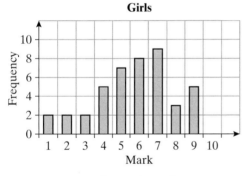

Boys

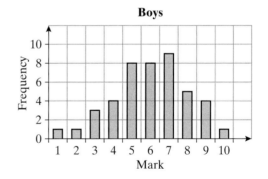

    (a) How many girls scored 7 in the test?
    (b) How many boys scored 4 in the test?
    (c) What was the modal score for the girls?
    (d) What was the modal score for the boys?
    (e) What was the range of the scores for the girls?
    (f) Did girls or boys do better in this test?
        Give a reason for your answer.

3 One measure of the health of a country is its Infant Mortality Rate (IMR).
The bar charts below show the IMR of five developed and five undeveloped countries.

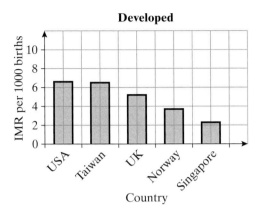

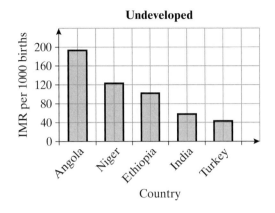

(a) Approximately, what is the IMR of the UK?
(b) Approximately, what is the IMR of Angola?
(c) Approximately, what is the mean IMR for the undeveloped countries?
(d) Approximately, what is the mean IMR for the developed countries?
(e) Approximately, what is the range of the IMRs for all 10 countries?
(f) Which group of countries is healthier?
(g) Draw a bar chart to show the data for all ten countries.

Exercise 14.1H

Write each of these numbers as a product of its prime factors.

1 14

2 16

3 28

4 35

5 42

6 49

7 108

8 156

9 225

10 424

Exercise 14.2H

For each of these pairs of numbers
- write the numbers as products of their prime factors.
- state the highest common factor.
- state the lowest common multiple.

1 6 and 8

2 8 and 18

3 15 and 25

4 36 and 48

5 25 and 55

6 33 and 55

7 54 and 72

8 30 and 40

9 45 and 63

10 24 and 50

Exercise 14.3H

Work out these.

1 2×3

2 -5×8

3 -6×-2

4 -4×6

5 5×-7

6 -3×7

7 -4×-5

8 $28 \div -7$

9 $-25 \div 5$

10 $-20 \div 4$

11 $24 \div 6$

12 $-15 \div -3$

13 $-35 \div 7$

14 $64 \div -8$

15 $27 \div -9$

16 $3 \times 6 \div -9$

17 $-42 \div -7 \times -3$

18 $5 \times 6 \div -10$

19 $-9 \times 4 \div -6$

20 $-5 \times 6 \times -4 \div -8$

Exercise 14.4H

1 Write down the value of each of these.

(a) 1^2 (b) 13^2 (c) $\sqrt{64}$

(d) $\sqrt{196}$ (e) 3^3 (f) 5^3

(g) $\sqrt[3]{8}$ (h) $\sqrt[3]{64}$

2 A cube has sides of length 4 cm.
What is its volume?

3 Find the square of each of these numbers.

(a) 20   (b) 42   (c) 5.1

(d) 60   (e) 0.9

4 Find the cube of each of these numbers.

(a) 7   (b) 3.5   (c) 9.4

(d) 20   (e) 100

5 Find the square roots of each of these numbers.
Where necessary, give your answer correct to
2 decimal places.

(a) 900   (b) 75   (c) 284

(d) 31 684   (e) 40 401

6 Find the cube root of each of these numbers.
Where necessary, give your answer correct to
2 decimal places.

(a) 729   (b) 144   (c) 9.261

(d) 4848   (e) 100 000

7 A square has an area of 80 cm².
What is the length of one side?
Give your answer correct to 2 decimal places.

Exercise 14.5H

1 Write down the reciprocal of each of these
numbers.

(a) 4   (b) 9   (c) 65

(d) 10   (e) 4.5

2 Write down the numbers of which these are the
reciprocals.

(a) $\frac{1}{6}$ (b) $\frac{1}{10}$ (c) $\frac{1}{25}$

(d) $\frac{1}{71}$ (e) $\frac{2}{15}$

3 Find the reciprocal of each of these numbers.
Give your answers as fractions or mixed
numbers.

(a) $\frac{3}{5}$ (b) $\frac{4}{9}$ (c) $2\frac{2}{5}$

(d) $5\frac{1}{3}$ (e) $\frac{3}{100}$

4 Find the reciprocal of each of these numbers.
Give your answers as decimals.

(a) 25   (b) 0.2   (c) 6.4

(d) 625   (e) 0.16

Algebra 2

Exercise 15.1H

Expand these.

1 $7(3a + 6b)$ 2 $5(2c + 3d)$

3 $4(3e - 5f)$ 4 $3(7g - 2h)$

5 $3(4i + 2j - 3k)$ 6 $3(5m - 2n + 3p)$

7 $6(4r - 3s - 2t)$ 8 $8(4r + 2s + t)$

9 $4(3u + 5v)$ 10 $6(4w + 3x)$

11 $2(5y + z)$ 12 $4(3y + 2z)$

13 $5(3v + 2)$ 14 $3(7 + 4w)$

15 $5(1 - 3a)$ 16 $3(8g - 5)$

17 $p(r - s)$ 18 $a(3 + b)$

19 $x(4 + 5y)$ 20 $y(7z - 1)$

21 $c(a + 2b - 3d)$

Exercise 15.2H

Expand the brackets and simplify these.

1 (a) $3(4a + 5) + 2(3a + 4)$
   (b) $5(4b + 3) + 3(2b + 1)$
   (c) $2(3 + 6c) + 4(5 + 7c)$

2 (a) $2(4x + 5) + 3(5x - 2)$
   (b) $4(3y + 2) + 5(3y - 2)$
   (c) $3(4 + 7z) + 2(3 - 5z)$

3 (a) $4(4s + 3t) + 5(2s + 3t)$
   (b) $3(4v + 5w) + 2(3v + 2w)$
   (c) $6(2x + 5y) + 3(4x + 2y)$
   (d) $2(5v + 4w) + 3(2v + w)$

4 (a) $5(2n + 5p) + 4(2n - 5p)$
   (b) $3(4q + 6r) + 5(2q - 3r)$
   (c) $7(3d + 2e) + 5(3d - 2e)$
   (d) $5(3f + 8g) + 4(3f - 9g)$
   (e) $4(5h - 6j) - 6(2h - 5j)$
   (f) $4(5k - 6m) - 3(2k - 5m)$

Exercise 15.3H

Factorise these.

1 (a) $8x + 20$ (b) $3x + 6$
   (c) $9x - 12$ (d) $5x - 30$

2 (a) $16 + 8x$ (b) $9 + 15x$
   (c) $12 - 16x$ (d) $8 - 12x$

3 (a) $4x^2 + 16x$ (b) $6x^2 + 30x$
   (c) $8x^2 - 20x$ (d) $9x^2 - 15x$

Exercise 15.4H

Simplify each of the following, writing your answer using index notation.

1 (a) $7 \times 7 \times 7 \times 7 \times 7$
   (b) $3 \times 3 \times 3 \times 3 \times 3$

2 (a) $d \times d \times d \times d \times d \times d \times d$
   (b) $m \times m \times m \times m \times m \times m$

3 (a) $a \times a \times a \times a \times b \times b$
   (b) $c \times c \times c \times c \times d \times d \times d \times d$

4 (a) $2x \times 3y \times 6z$
   (b) $r \times 2s \times 3t \times 4s \times 5r$

Exercise 16.1H

1 For a project, Rebecca recorded the ages of 100 cars as they passed the school gates one morning. Here are her results.

Age (a years)	Frequency
$0 \leqslant a < 2$	16
$2 \leqslant a < 4$	23
$4 \leqslant a < 6$	24
$6 \leqslant a < 8$	17
$8 \leqslant a < 10$	12
$10 \leqslant a < 12$	7
$12 \leqslant a < 14$	1

(a) Draw a frequency diagram to show these data.
(b) Which of the intervals is the modal class?

2 The manager of a leisure centre recorded the weights of 120 men.
Here are the results.

Weight (w kg)	Frequency
$60 \leqslant w < 65$	4
$65 \leqslant w < 70$	18
$70 \leqslant w < 75$	36
$75 \leqslant w < 80$	50
$80 \leqslant w < 85$	10
$85 \leqslant w < 90$	2

(a) Draw a frequency diagram to represent these data.
(b) Which of the intervals is the modal class?
(c) Which of the intervals contains the median value?

 3 This frequency diagram shows the times taken by a group of girls to run a race.

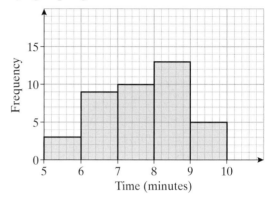

(a) How many girls took longer than 9 minutes?
(b) How many girls took part in the race?
(c) What percentage of the girls took less than 7 minutes?
(d) What is the modal finishing time?
(e) Use the diagram to draw up a grouped frequency table like those in questions 1 and 2.

Exercise 16.2H

1 The table shows the heights of 40 plants.

Height (h cm)	Frequency
$3 \leqslant h < 4$	1
$4 \leqslant h < 5$	7
$5 \leqslant h < 6$	10
$6 \leqslant h < 7$	12
$7 \leqslant h < 8$	8
$8 \leqslant h < 9$	2

Draw a frequency polygon to show these data.

2 The table shows the times taken for a group of children to get from home to school.

Time (*t* mins)	Frequency
$0 \leqslant t < 5$	3
$5 \leqslant t < 10$	15
$10 \leqslant t < 15$	27
$15 \leqslant t < 20$	34
$20 \leqslant t < 25$	19
$25 \leqslant t < 30$	2

Draw a frequency polygon to show these data.

 3 The ages of all of the people under 70 in a small village were recorded in 1985 and 2005.
The results are given in the table below.

Age (*a* years)	Frequency 1985	Frequency 2005
$0 \leqslant a < 10$	85	50
$10 \leqslant a < 20$	78	51
$20 \leqslant a < 30$	70	78
$30 \leqslant a < 40$	53	76
$40 \leqslant a < 50$	40	62
$50 \leqslant a < 60$	28	64
$60 \leqslant a < 70$	18	56

(a) On the same grid, draw a frequency polygon for each year.

(b) Use the diagram to compare the distribution of ages in the two years.

Exercise 16.3H

1 As part of a survey, Emma measured the heights, in centimetres, of the 50 teachers in her school. Here are her results.

168	194	156	167	177
180	188	172	170	169
174	178	186	174	166
165	159	173	185	162
163	174	180	184	173
182	161	176	170	169
178	175	172	179	173
162	177	176	184	191
181	165	163	185	178
175	182	164	179	168

Construct a stem-and-leaf diagram to show these heights.

2 A group of students took a maths test. Here are their marks.

51	94	56	45	70	67	69	49	55	71
52	73	64	60	71	58	64	35	31	81
41	67	64	73	82	57	75	33	88	79
66	52	48							

(a) Construct a stem-and-leaf diagram to show these marks.

(b) The pass mark was 40.
How many students had to do a re-test?

(c) What is the modal mark?

(d) What is the median mark?

3 The data shows the length of the throws, in metres, in a school shot putt competition.

4.4	5.4	8.5	9.2	7.3	5.7	9.9	9.6	7.4
9.1	7.5	7.0	8.3	4.9	5.7	6.4	6.7	7.3
8.2	9.0	5.2	7.0	8.9	9.1	5.2	6.4	7.3
8.2	5.9	5.0	6.5	7.9	8.2	8.9	5.3	5.2

(a) Construct a stem-and-leaf diagram to show these lengths.

(b) How many people took part in the competition?

(c) Find the median length.

Exercise 17.1H

Solve these equations.

1 $2x - 3 = 7$

2 $2x + 2 = 8$

3 $2x - 9 = 3$

4 $3x - 2 = 7$

5 $6x + 2 = 26$

6 $3x + 2 = 17$

7 $4x - 5 = 3$

8 $4x + 2 = 8$

9 $2x - 7 = 10$

10 $5x + 12 = 7$

11 $x^2 + 3 = 19$

12 $x^2 - 2 = 7$

13 $y^2 - 1 = 80$

14 $11 - 3x = 2$

Exercise 17.2H

Solve these equations.

1 $3(x - 2) = 18$

2 $2(1 + x) = 8$

3 $3(x - 5) = 6$

4 $2(x + 3) = 10$

5 $5(x - 2) = 15$

6 $2(x + 3) = 10$

7 $5(x - 4) = 20$

8 $4(x + 1) = 16$

9 $2(x - 7) = 8$

10 $3(2x + 3) = 18$

11 $5(2x - 3) = 15$

12 $2(3x - 2) = 14$

13 $5(2x - 3) = 40$

14 $4(x - 3) = 6$

15 $3(2x + 3) = 18$

16 $4(3y - 7) = 16$

17 $5(4x + 2) = 15$

18 $2(5x - 3) = 4$

Exercise 17.3H

Solve these equations.

1 $5x - 1 = 3x + 5$

2 $5x + 1 = 2x + 13$

3 $7x - 2 = 2x + 8$

4 $6x + 1 = 4x + 21$

5 $9x - 10 = 4x + 5$

6 $5x - 8 = 3x - 6$

7 $6x + 2 = 10 - 2x$

8 $2x - 10 = 5 - 3x$

9 $15 + 3x = 2x + 18$

10 $2x - 5 = 4 - x$

11 $3x - 2 = x + 7$

12 $x - 1 = 2x - 6$

13 $2x - 4 = 2 - x$

14 $9 - x = x + 5$

15 $5x + 3 = 2(x - 9)$

16 $3(x + 4) = x + 2$

Exercise 17.4H

Solve these equations.

1 $\frac{x}{2} = 7$

2 $\frac{x}{5} - 2 = 1$

3 $\frac{x}{4} + 5 = 8$

4 $\frac{x}{3} - 5 = 5$

5 $\frac{x}{6} + 3 = 4$

6 $\frac{x}{5} + 1 = 4$

7 $\frac{x}{8} - 3 = 9$

8 $\frac{x}{4} + 1 = 3$

9 $\frac{x}{7} + 5 = 6$

10 $\frac{x}{4} + 5 = 4$

Chapter 18 | Ratio and proportion

Exercise 18.1H

1 Write each of these ratios in its lowest terms.
(a) 8 : 6   (b) 20 : 50   (c) 35 : 55
(d) 8 : 24 : 32   (e) 15 : 25 : 20

2 Write each of these ratios in its lowest terms.
(a) 200 g : 500 g
(b) 60p : £3
(c) 1 minute : 25 seconds
(d) 2 m : 80 cm
(e) 500 g : 3 kg

3 A bar of brass contains 400 g of copper and 200 g of zinc.
Write the ratio of copper to zinc in its lowest terms.

4 Teri, Jannae and Abi receive £200, £350 and £450 respectively as their dividends in a joint investment.
Write the ratio of their dividends in its lowest terms.

5 Three saucepans hold 500 ml, 1 litre and 2.5 litres respectively.
Write the ratio of their capacities in its lowest terms.

Exercise 18.2H

1 Write each of these ratios in the form 1 : n.
(a) 2 : 10   (b) 5 : 30
(c) 2 : 9   (d) 4 : 9
(e) 50 g : 30 g   (f) 15p : £3
(g) 25 cm : 6 m   (h) 20 : 7
(i) 4 mm : 1 km

2 On a map a distance of 12 mm represents a distance of 3 km.
What is the scale of the map in the form 1 : n?

3 A picture is enlarged on a photocopier from 25 mm wide to 15 cm wide.
What is the ratio of the picture to the enlargement in the form 1 : n?

Exercise 18.3H

1 A photo is enlarged in the ratio 1 : 5.
(a) The length of the small photo is 15 cm.
What is the length of the large photo?
(b) The width of the large photo is 45 cm.
What is the width of the small photo?

2 To make a dressing for her lawn, Rupinder mixes loam and sand in the ratio 1 : 3.
(a) How much sand should she mix with two buckets of loam?
(b) How much loam should she mix with 15 buckets of sand?

3 To make mortar, Fred mixes 1 part cement with 5 parts sand.
(a) How much sand does he mix with 500 g of cement?
(b) How much cement does he mix with 4.5 kg of sand?

4 A rectangular picture is 6 cm wide.
It is enlarged in the ratio 1 : 4.
How wide is the enlargement?

5 The Michelin motoring atlas of France has a scale of 1 cm to 2 km.
(a) On the map the distance between Metz and Nancy is 25 cm.
How far is the actual distance between the two towns?
(b) The actual distance between Caen and Falaise is 33 km.
How far is this on the map?

6 Graham is making pastry.
To make enough for five people he uses 300 g of flour.
How much flour should he use for eight people?

7 To make a solution of a chemical a scientist mixes 2 parts chemical with 25 parts water.
(a) How much water should he mix with 10 ml of chemical?
(b) How much chemical should he mix with 1 litre of water?

8 The ratio of the sides of two rectangles is 2 : 5.
   (a) The length of the small rectangle is 4 cm. How long is the big rectangle?
   (b) The width of the big rectangle is 7.5 cm. How wide is the small rectangle?

9 Jason mixes 3 parts black paint with 4 parts white paint to make dark grey paint.
   (a) How much white paint does he mix with 150 ml of black paint?
   (b) How much black paint does he mix with 1 litre of white paint?

10 In an election the number of votes was shared between the Labour, Conservatives and other parties in the ratio 5 : 4 : 2.
   Labour received 7500 votes.
   (a) How many votes did the Conservatives receive?
   (b) How many votes did the other parties receive?

Exercise 18.4H

1 Share £40 in the ratio 3 : 5.

2 Paint is mixed in the ratio 2 parts black paint to 3 parts white paint to make 10 litres of grey paint.
   (a) How much black paint is used?
   (b) How much white paint is used?

3 A metal alloy is made up of copper, iron and nickel in the ratio 3 : 4 : 2.
   How much of each metal is there in 450 g of the alloy?

4 Inderjit worked 6 hours one day.
   The time he spent on filing, writing and computing is in the ratio 2 : 3 : 7.
   How long did he spend computing?

5 Daisy and Emily invested £5000 and £8000 respectively in a business venture.
   They agreed to share the profits in the ratio of their investment.
   Emily received £320.
   What was the total profit?

6 Shahida spends her pocket money on sweets, magazines and clothes in the ratio 2 : 3 : 7.
   She receives £15 a week.
   How much does she spend on sweets?

7 In a questionnaire the three possible answers are 'Yes', 'No' and 'Don't know'.
   The answers from a group of 456 people are in the ratio 10 : 6 : 3.
   How many 'Don't knows' are there?

8 Iain and Stephen bought a house between them in Spain.
   Iain paid 60% of the cost and Stephen 40%.
   (a) Write the ratio of the amounts they paid in its lowest terms.
   (b) The house cost 210 000 euros. How much did each pay?

Exercise 18.5H

1 An 80 g bag of Munchoes costs 99p and a 200 g bag of Munchoes costs £2.19.
   Show which is the better value.

2 Baxter's lemonade is sold in 2 litre bottles for £1.29 and in 3 litre bottles for £1.99.
   Show which is the better value.

3 Butter is sold in 200 g tubs for 95p and in 450 g packets for £2.10.
   Show which is the better value.

4 Fruit yogurt is sold in packs of 4 tubs for 79p and in packs of 12 tubs for £2.19.
   Show which is the better value.

5 There are two packs of minced meat on the reduced price shelf at the supermarket, a 1.8 kg pack reduced to £2.50 and a 1.5 kg pack reduced to £2.
   Show which is the better value.

6 Smoothie shaving gel costs £1.19 for the 75 ml bottle and £2.89 for the 200 ml bottle.
   Show which is better value.

7 A supermarket sells cans of cola in two different sized packs: a pack of 12 cans costs £4.30 and a pack of 20 cans costs £7.25.
   Which pack gives the better value?

8 Sudso washing powder is sold in three sizes: 750 g for £3.15, 1.5 kg for £5.99 and 2.5 kg for £6.99.
   Which size gives the best value?

Exercise 19.1H

1 For each of these sets of data
   (i)   find the mode.
   (ii) find the median.
   (iii) find the range.
   (iv) calculate the mean.

(a)

Score on biased dice	Number of times thrown
1	52
2	46
3	70
4	54
5	36
6	42
Total	300

(b)

Number of drawing pins in a box	Number of boxes
98	5
99	14
100	36
101	28
102	17
103	13
104	7
Total	120

(c)

Number of snacks per day	Frequency
0	23
1	68
2	39
3	21
4	10
5	3
6	1

(d)

Number of letters received on Monday	Frequency
0	19
1	37
2	18
3	24
4	12
5	5
6	2
7	3

2 Gift tokens cost £1, £5, £10, £20 or £50 each.
The frequency table shows the numbers of each value of gift token sold in one bookstore on a Saturday.

Price of gift token (£)	1	5	10	20	50
Number of tokens sold	12	34	26	9	1

Calculate the mean value of gift token bought in the bookstore that Saturday.

4 A sample of people were asked how many visits to the cinema they had made in one month.
None of those asked had made more than eight visits to the cinema.
The table shows the data.

Number of visits	0	1	2	3	4	5	6	7	8
Frequency	136	123	72	41	18	0	5	1	4

Calculate the mean number of visits to the cinema.

Exercise 19.2H

1 For each of these sets of data, calculate an estimate of
(i) the range.
(ii) the mean.

(a)

Number of trains arriving late each day (x)	Number of days (f)
0–4	19
5–9	9
10–14	3
15–19	0
20–24	1
Total	32

(b)

Number of weeds per square metre (x)	Number of square metres (f)
0–14	204
15–29	101
30–44	39
45–59	13
60–74	6
75–89	2

(c)

Number of books sold (x)	Frequency (f)
60–64	3
65–69	12
70–74	23
75–79	9
80–84	4
85–89	1

(d)

Number of days absent (x)	Frequency (f)
0–3	13
4–7	18
8–11	9
12–15	4
16–19	0
20–23	1
24–27	3

2 The table gives the number of sentences per chapter in a book.
(a) What is the modal class?
(b) In which class is the median number of sentences?
(c) Calculate an estimate of the mean number of sentences.

Number of sentences (x)	Frequency (f)
100–124	1
125–149	9
150–174	8
175–199	5
200–224	2

3 A group of students were asked to estimate the number of beans in a jar.
The results of their estimates are summarised in the table.
Calculate an estimate of the mean number of beans estimated by these students.

Estimated number of beans (x)	Frequency (f)
300–324	9
325–349	26
350–374	52
375–399	64
400–424	83
425–449	57
450–474	18
475–499	5

Exercise 19.3H

1 For each of these sets of data, calculate an estimate, to the nearest whole number, of
   (i) the range.
   (ii) the mean.

(a)

Height of sunflower in centimetres (x)	Number of plants (f)
$100 \leqslant x < 110$	6
$110 \leqslant x < 120$	13
$120 \leqslant x < 130$	35
$130 \leqslant x < 140$	29
$140 \leqslant x < 50$	16
$150 \leqslant x < 160$	11
Total	110

(b)

Weight of egg in grams (x)	Number of eggs (f)
$20 \leqslant x < 25$	9
$25 \leqslant x < 30$	16
$30 \leqslant x < 35$	33
$35 \leqslant x < 40$	48
$40 \leqslant x < 45$	29
$45 \leqslant x < 50$	15
Total	150

(c)

Length of green bean in millimetres (x)	Frequency (f)
$60 \leqslant x < 80$	12
$80 \leqslant x < 100$	21
$100 \leqslant x < 120$	46
$120 \leqslant x < 140$	27
$140 \leqslant x < 160$	14
Total	120

(d)

Time to complete race in minutes (x)	Frequency (f)
$54 \leqslant x < 56$	1
$56 \leqslant x < 58$	4
$58 \leqslant x < 60$	11
$60 \leqslant x < 62$	6
$62 \leqslant x < 64$	2
$64 \leqslant x < 66$	1
Total	25

2 For each of these sets of data
   (i) write down the modal class.
   (ii) calculate an estimate of the mean.

(a)

Height of shrub in metres (x)	Number of shrubs (f)
$0.3 \leqslant x < 0.6$	57
$0.6 \leqslant x < 0.9$	41
$0.9 \leqslant x < 1.2$	36
$1.2 \leqslant x < 1.5$	24
$1.5 \leqslant x < 1.8$	15

(b)

Weight of plum in grams (x)	Number of plums (f)
$20 \leqslant x < 30$	6
$30 \leqslant x < 40$	19
$40 \leqslant x < 50$	58
$50 \leqslant x < 60$	15
$60 \leqslant x < 70$	4

(c)

Length of journey in minutes (x)	Frequency (f)
$20 \leqslant x < 22$	6
$22 \leqslant x < 24$	20
$24 \leqslant x < 26$	38
$26 \leqslant x < 28$	47
$28 \leqslant x < 30$	16
$30 \leqslant x < 32$	3

(d)

Speed of car in miles per hour (x)	Frequency (f)
$25 \leqslant x < 30$	4
$30 \leqslant x < 35$	29
$35 \leqslant x < 40$	33
$40 \leqslant x < 45$	6
$45 \leqslant x < 50$	2
$50 \leqslant x < 55$	1

3 The table shows the monthly wages of the workers in an office.
   (a) What is the modal class?
   (b) In which class is the median wage?
   (c) Calculate an estimate of the mean wage.

Wages in £ (x)	Frequency (f)
$500 \leqslant x < 1000$	3
$1000 \leqslant x < 1500$	14
$1500 \leqslant x < 2000$	18
$2000 \leqslant x < 2500$	5

4 The table shows the length, in seconds, of 100 calls made from a mobile phone.
   Calculate an estimate of the mean length of a call.

Length of call in seconds (x)	Frequency (f)
$0 \leqslant x < 30$	51
$30 \leqslant x < 60$	25
$60 \leqslant x < 90$	13
$90 \leqslant x < 120$	7
$120 \leqslant x < 150$	4

5 The table shows the prices paid for greetings cards sold in one day by a card shop.
   Calculate an estimate of the mean price, in pence, paid for a greetings card that day.

Price of greetings card in pence (x)	Frequency (f)
$75 \leqslant x < 100$	23
$100 \leqslant x < 125$	31
$125 \leqslant x < 150$	72
$150 \leqslant x < 175$	59
$175 \leqslant x < 200$	34
$200 \leqslant x < 225$	11
$225 \leqslant x < 250$	5

Pythagoras' theorem

Exercise 20.1H

For each of these diagrams, find the area of the third square.

1

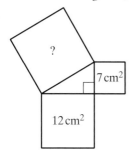

2

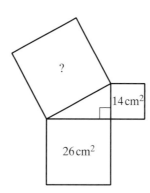

3

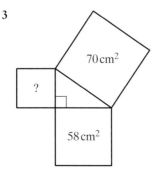

4

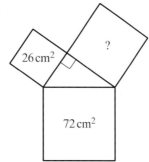

Exercise 20.2H

1 For each of these triangles, find the length of the hypotenuse, marked x.
Where the answer is not exact, give your answer correct to 2 decimal places.

(a)

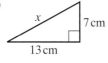

(b)

(c)

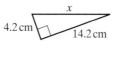

(d)

2 For each of these triangles, find the length of the shorter side, marked x.
Where the answer is not exact, give your answer correct to 2 decimal places.

(a)

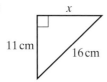

(b)

(c)

(d)

3 Ann can walk home from school along two roads or along a path across a field.
How much shorter is her journey if she takes the path across the field?

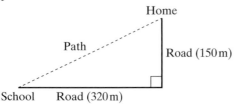

4 This network is made of wire.
What is the total length of wire?

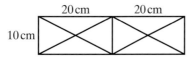

Exercise 20.3H

Work out whether or not each of these triangles is right-angled.
Show your working.

1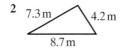

2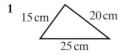

3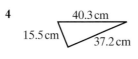

4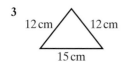

5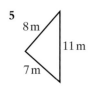

6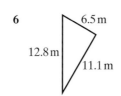

7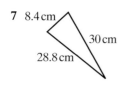

8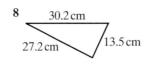

Exercise 20.4H

1 For each of the line segments in the diagram
(i) find the coordinates of the midpoint.
(ii) find the length.
Where the answer is not exact, give your answer to 2 decimal places.

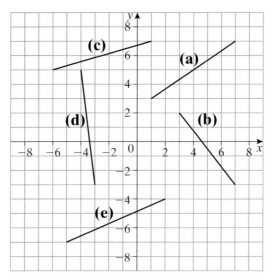

2 For the line segments joining each of the pairs of points below
(i) find the coordinates of the midpoint.
(ii) find the length.
Where the answer is not exact, give your answer to 2 decimal places.
(a) A(3, 7) and B(−5, 7)
(b) C(2, 1) and D(8, 5)
(c) E(3, 7) and F(8, 2)
(d) G(1, 6) and H(9, 3)
(e) I(−7, 1) and J(3, 6)
(f) K(−5, −6) and L(−7, −3)

Chapter 21

Planning and collecting

Exercise 21.1H

1 State whether the following would give primary data or secondary data.
   (a) Weighing packets of sweets
   (b) Using bus timetables
   (c) Looking up holiday prices on the internet
   (d) A GP entering data for a new patient on his records after seeing the patient

2 Lisa is doing a survey and has written this question.

   > What colour is your hair?
   >
   > Black ☐   Brown ☐   Blonde ☐

   (a) Give a reason why this question is unsuitable.
   (b) Write a better version.

3 Steve is doing a survey about his local sports facilities.
   Here is one of his questions.

   > How much do you enjoy doing sport?
   >   1   2   3   4   5

   (a) Give a reason why this question is unsuitable.
   (b) Write a better version.

4 Mia is doing a survey about school lunches. She gives out questionnaires to the first 30 people in the queue for lunch.
   (a) Why is this likely to give a biased sample?
   (b) Describe a better method of obtaining a sample for her survey.

5 Here is one of Mia's questions.

   > Don't you agree that we don't have enough salads on the menu?

   (a) Give a reason why this question is unsuitable.
   (b) Write a better version.

6 A survey is to be done about school students' earnings and pocket money.
   Write five suitable questions which could be included in such a survey.

Exercise 22.1H

1 Look at this sequence of patterns.
The first four patterns in the sequence have been drawn.

(a) Describe the position-to-term rule for this sequence.
(b) How many circles are there in the 100th pattern?

 2 Look at this sequence of matchstick patterns.

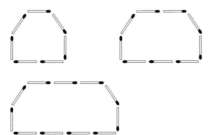

(a) Copy and complete this table.

Pattern number	Number of matchsticks
1	
2	
3	
4	
5	

(b) What patterns can you see in the numbers?
(c) Find the number of matchsticks in the 50th pattern.

 3 Here is a sequence of star patterns.

(a) Draw the next pattern in the sequence.
(b) Without drawing the pattern, find the number of stars in the 8th pattern.
Explain how you found your answer.

4 The numbers in a sequence are given by this rule:
Multiply the position number by 7, then subtract 10.
(a) Show that the first term of the sequence is −3.
(b) Find the next four terms of the sequence.

5 Find the first four terms of the sequences with these nth terms.
(a) $10n$ (b) $8n + 2$

6 Find the first five terms of the sequences with these nth terms.
(a) n^2 (b) $2n^2$ (c) $5n^2$

 7 The first term of a sequence is 3.
The general rule for the sequence is multiply a term by 3 to get to the next term.
Write down the first five terms of the sequence.

 8 For a sequence, $T_1 = 12$ and $T_{n+1} = T_n - 5$.
Write down the first four terms of this sequence.

 9 Draw suitable patterns to represent this sequence.

1, 4, 7, 10, ...

 10 Draw suitable patterns to represent this sequence.

1 × 1, 3 × 3, 5 × 5, 7 × 7, ...

Exercise 22.2H

1 Find the *n*th term for each of these sequences.
  (a) 10, 13, 16, 19, 22, ...
  (b) 0, 1, 2, 3, 4, ...
  (c) −3, −1, 1, 3, 5, ...

2 Find the *n*th term for each of these sequences.
  (a) 25, 20, 15, 10, 5, ...
  (b) 4, 2, 0, −2, −4, ...
  (c) 3, 2, 1, 0, −1, ...

3 Which of these sequences are linear?
  Find the next two terms of each of the sequences that are linear.
  (a) 2, 5, 10, 17, ...
  (b) 2, 5, 8, 11, ...
  (c) 1, 3, 6, 10, ...
  (d) 12, 8, 4, 0, −4, ...

 4 (a) Write down the first five terms of the sequence with *n*th term 100*n*.
  (b) Compare your answers with this sequence.

    99, 199, 299, 399, ...

  Write down the *n*th term of this sequence.

5 A mail-order shirt company charges £25 per shirt, plus an overall delivery charge of £3.
  (a) Copy and complete the table.

Number of shirts	1	2	3
Cost in £			

  (b) Write an expression for the cost, in pounds, of *n* shirts.
  (c) Paul pays £128 for shirts.
    How many does he buy?

 6 (a) Write down the first five terms of the sequence with *n*th term n^2.
  (b) Compare your answers with this sequence.

    0, 3, 8, 15, 24, ...

  Write down the *n*th term of this sequence.

7 The *n*th triangular number is $\dfrac{n(n+1)}{2}$.
  Find the 60th triangular number.

 8 The *n*th term of a sequence is 2^n.
  (a) Write down the first five terms of this sequence.
  (b) Describe the sequence.

 9 (a) Write down the first five cube numbers.
  (b) Compare this sequence with the sequence of the cube numbers.

    3, 10, 29, 66, 127, ...

  Use what you notice to write down the *n*th term of this sequence.
  (c) Find the 10th term of this sequence.

10 (a) Compare this sequence with the sequence of square numbers.

    5, 20, 45, 80, 125, ...

  Use what you notice to write down the *n*th term of this sequence.
  (b) Find the 20th term of this sequence.

Exercise 23.1H

1 Two points, A and B, are 7 cm apart.
   Construct the locus of points that are equidistant
   from A and B.

2 A badger will never go further than 3 miles from
   its home.
   Construct a scale diagram to show the regions
   where the badger might go looking for food.

3 Draw an equilateral triangle, ABC, of side 6 cm.
   Shade the region of points inside the triangle
   which are nearer to AB than to AC.

4 A rectangular garden measures 8 m by 6 m.
   A fence is built from F, at a right angle across the
   garden.
   Draw a scale diagram and construct the line of
   the fence.

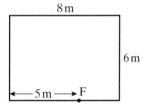

5 Draw a square, ABCD, of side 5 cm.
   Construct the locus of points, inside the square,
   which are more than 3 cm from A.

6 Zeke is walking across a field.
   He notices a bull starting to chase after him.
   He runs the shortest distance to the hedge.
   Make a larger copy of the diagram and construct
   the path that Zeke should run.

• Zeke

7 Draw a rectangle, PQRS, with sides PQ = 7 cm
   and QR = 5 cm.
   Shade the region of points within PQRS that are
   closer to P than to Q.

8 Draw an angle of 80°.
   Construct the bisector of the angle.

9 An office is a rectangle measuring 16 m by 12 m.
   There are two electricity points in the office at
   opposite corners of the room.
   The vacuum cleaner has a wire 10 m long.
   Make a scale drawing to show how much of the
   room can be cleaned.

10 Make another scale drawing of the office in
    question **9**.
    Shade the locus of points which are equidistant
    from the two electricity points.
    Use this locus to work out the length of wire
    needed for the vacuum cleaner to reach
    everywhere in the office.

Exercise 23.2H

 1 Draw a point and label it P.
Construct the locus of points that are less than 4 cm from P or more than 6 cm from P.

 2 A rectangular garden measures 20 m by 12 m.
A tree is to be planted so that it is more than 4 m from each corner of the garden.
Make a scale drawing to find the area where the tree can be planted.

 3 Two points, A and B, are 5 cm apart.
Find the region that is less than 3 cm from A and more than 4 cm from B.

 4 Make an accurate drawing of a triangle, PQR, where PQ = 6 cm, P = 40° and Q = 35°.
Find the point X, which is 2 cm from R and equidistant from P and Q.

 5 Two coastguard stations, A and B, are 20 km apart on a straight coastline.
The coastguard at A knows that a ship is within 15 km of him.
The coastguard at B knows that the same ship is within 10 km of him.
Make a scale drawing to show the region where the ship could be.

 6 Draw two lines, each 6 cm long, joined to form a right angle.
Draw the region of points which are less than 3 cm from these lines.

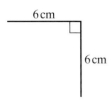

 7 Two points, P and Q, are 7 cm apart.
Find the points which are the same distance from P and Q and are also within 5 cm of Q.

 8 The diagram shows three coastguard stations, C, D and E.
A ship is within 25 km of C and closer to DE than DC.
Make a scale drawing and show the region where the ship could be.

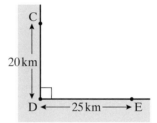

 9 A garden is a rectangle, ABCD, with AB = 5 m and BC = 3 m.
A new flower bed is to be made in the garden.
It must be more than 2 m from A and less than 1.5 m from CD.
Make a scale drawing to show where the flower bed could be.

 10 EFG is a triangle with EF = 6 cm, FG = 8 cm and EG = 10 cm.
Construct the perpendicular from F to EG.
Indicate the points on this line that are more than 7 cm from G.

Chapter 24 Rearranging formulae

Exercise 24.1H

1 Rearrange each of these formulae to make the letter in brackets the subject.
   (a) $a = b + c$ (b)
   (b) $a = 3x - y$ (x)
   (c) $a = b + ct$ (t)
   (d) $F = 2(q + p)$ (q)
   (e) $x = 2y - 3z$ (y)
   (f) $P = \dfrac{3 + 4n}{5}$ (n)

2 The formula for the circumference of a circle is
   $C = \pi d$.
   Rearrange the formula to make d the subject.

3 Rearrange the formula $A = \dfrac{3ab}{2n}$ to make
   (a) a the subject.
   (b) n the subject.

4 The formula for finding the perimeter of a rectangle is $P = 2(a + b)$, where P is the perimeter, a is the length and b is the width of the rectangle.
   Rearrange the formula to make a the subject.

5 The formula $y = mx + c$ is the equation of a straight line.
   Rearrange it to find m in terms of x, y and c.

6 The surface area of a sphere is given by the formula $A = 4\pi r^2$.
   Rearrange the formula to make r the subject.

7 The formula for the volume of this prism is
   $V = \dfrac{\pi r^2 h}{4}$.

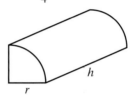

   (a) Find V when $r = 2.5$ and $h = 7$.
   (b) (i) Rearrange the formula to make r the subject.
       (ii) Find r when $V = 100$ and $h = 10$.

8 The formula for the surface area of a closed cylinder is $S = 2\pi r(r + h)$.
   Rearrange the formula to make h the subject.

Unit B Contents

Chapter 1 Working with numbers

Exercise 1.1H

1 Put these whole numbers in order of size, smallest first.
   (a) 782, 2807, 1872, 287, 87, 708
   (b) 439, 3047, 8372, 48, 843, 4389
   (c) 1260, 10086, 80060, 100000, 806

2 Here are five digits: 5, 1, 6, 4, 8.
   Using each digit only once, write down
   (a) the smallest whole number that can be made with the five digits.
   (b) the largest whole number that can be made with the five digits.

3 Each of these groups of numbers are in order of size.
   Write each group out with an extra number between each one.
   (a) 3   6   17   27   47   99   121
   (b) 607   670   687   697   700   704   707   717

Exercise 1.2H

1 How many do you need to add to these numbers to make 10?
   (a) 2   (b) 9   (c) 7

2 How many do you need to add to these numbers to make 100?
   (a) 88   (b) 64   (c) 19

3 Write down the complements to 10 of these numbers.
   (a) 5   (b) 3   (c) 1

4 Write down the complements to 100 of these numbers.
   (a) 85   (b) 35   (c) 2

5 (a) What is the complements to 10 of 8?
   (b) What is 41 − 8?
   (c) Add the complement to 10 of 8 to 31?
   (d) What can you say about your answers to parts (b) and (c)?

6 (a) What is the complement to 100 of 58?
   (b) What is 140 − 58?
   (c) Add the complement to 100 of 58 to 40.
   (d) What can you say about your answers to parts (b) and (c)?

7 Use complements to work out these.
   (a) 34 − 7
   (b) 64 − 9
   (c) 156 − 67

Exercise 1.3H

Work out these.

1 2×3

2 3×2

3 7×6

4 6×7

5 7×2

6 4×8

7 9×9

8 5×4

9 3×3

10 8×3

11 9×6

12 5×8

13 3×7

14 6×4

15 9×10

16 8×8

17 10×3

18 5×7

19 9×7

20 8×9

Exercise 1.4H

Work out these.

1 $8 \div 2$

2 $3\overline{)24}$

3 $20 \div 5$

4 $24 \div 4$

5 $35 \div 5$

6 $7\overline{)14}$

7 $18 \div 6$

8 $42 \div 7$

9 $21 \div 3$

10 $9\overline{)72}$

11 $30 \div 6$

12 $8\overline{)72}$

13 $48 \div 6$

14 $7\overline{)28}$

15 $54 \div 9$

16 $2\overline{)12}$

17 $27 \div 3$

18 $60 \div 6$

19 $56 \div 7$

20 $8\overline{)64}$

Exercise 1.5H

1 Use mental methods to work out these.
 (a) $157 + 97$ **(b)** $254 + 728$
 (c) $415 - 87$ **(d)** $231 - 63$
 (e) $815 + 146$ **(f)** $446 - 158$
 (g) $403 - 176$ **(h)** $854 + 161$

2 Add 382 and 498.

3 What is 714 take away 345?

4 What is 257 plus 96?

5 Find 613 minus 234.

6 What is the sum of 471 and 219?

7 What is 682 subtract 594?

8 Find the total of 456 and 345.

9 Jack has 473 stamps. Jill has 529.
 How many stamps do they have in total?

10 A pack of paper contains 500 sheets.
 Sam uses 314 sheets from the pack.
 How many sheets are left in the pack?

11 The road distance from Bloxton to Conbridge is
 524 kilometres. Pete has driven 266 kilometres
 along the route already.
 How many more kilometres does he have to drive?

12 I buy a T-shirt for £5.95.
 How much change do I get from £10?

13 Andy has these numbers written on a piece of
 paper.
 125   438   253   341   169   221
 Andy claims that he can make exactly 1000 by
 adding just three of the numbers together.
 Show whether this is possible.

14 Jade has these numbers written on a piece of
 paper.
 167   366   215   152   278   653
 She claims that, if she adds the two smallest
 numbers together and the two largest numbers
 together, the difference between her two answers
 is exactly 700.
 Explain whether this is true.

Exercise 1.6H

1 Work out these.
 (a) $12.43 + 36.31$ **(b)** $76.02 + 11.69$
 (c) $34.98 + 42.37$ **(d)** $5.73 + 8.58$
 (e) $8.09 + 13.75$ **(f)** $62.96 + 70.36$

2 Work out these.
 (a) $37.84 - 15.43$ **(b)** $59.26 - 38.16$
 (c) $87.45 - 24.27$ **(d)** $37.12 - 28.67$
 (e) $63.92 - 27.36$ **(f)** $18.48 - 16.50$

3 Work out these.
(a) £3.95 + £7.03 + £12.84
(b) £25 + £3.64 + 84p + £6.09

4 Work out these.
(a) £20.00 − 85p
(b) £28 + £18.95 − £12.62

5 Work out these.
Give your answers in the larger unit.
(a) 2.4 m + 1.35 m + 85 cm
(b) 90 cm + 4.56 m + 1.28 m + 45 cm
(c) 2.4 m − 58 cm
(d) 3.75 m − 855 mm

6 A wall is to be 25 m long.
So far, the bricklayer has built 18.75 m.
How much more has to be built?

7 An antique dealer bought two pictures, one for £35.50 and the other for 85p.
He sold the two together for £90.
How much money did he make?

8 Two pieces measuring 90 cm each are cut from a length of wood 2.4 m long.
What length is left?

9 Sandra has £5 in her purse.
She spends £1.40 on a magazine and 89p on a drink.
How much does she have left?

10 Here is a price list from a café.

Drinks	£1.15
Sandwich	£1.89
Salad	40p
Crisps	65p
Cake	£1.70

(a) Miya orders a cheese sandwich with salad, a packet of crisps and a drink.
How much does this cost?
(b) Miya pays with a £5 note.
How much change should she receive?
(c) Tom has £3.50.
He wants a drink and a cake.
What else could he buy?

Exercise 1.7H

1 Work out these.
(a) £4.37 + £5.23
(b) £2.61 + £3.42
(c) £8.46 + £5.79
(d) £5.83 − £2.54
(e) £6.23 + £3.74 − £4.66

2 Work out these.
(a) 3.87 + 9.15
(b) 38.43 + 59.12
(c) 41.53 + 67.42

3 Work out these.
(a) 19.28 − 6.25
(b) 47.16 − 15.42
(c) 253.80 − 81.47

4 Work out these.
(a) £3.95 + 82p + £1.57
(b) £15.21 + 77p + £3.42 + 61p
(c) £63.84 + 90p + £8.51 + £91.20 + 47p
(d) £6.25 + 42p + 87p + £63.20

5 In the javelin, Amina throws 52.15 m and Raj throws 46.47 m.
Find the difference between the lengths of their throws.

6 The times for the first and last places in a show-jumping event were 47.82 seconds and 53.19 seconds.
Find the difference between these times.

7 Holly buys two scarves at £8.45 each and a pair of shoes at £35.99.
How much change does she get from £60?

Exercise 1.8H

1 Work out these.
(a) 3.6 × 4   (b) 5.8 × 6   (c) 0.5 × 0.7
(d) 2.3 × 6.4   (c) 4.3 × 1.8   (f) 14.2 × 5.8
(g) 7.9 × 6.2   (h) 21.3 × 4.3

2 Work out these.
(a) 23.6 ÷ 4   (b) 23.4 ÷ 3   (c) 16.2 ÷ 0.3
(d) 32.8 ÷ 0.4   (e) 51.1 ÷ 0.7   (f) 82.8 ÷ 1.8

3 Given that $74 \times 386 = 28\,564$, write down the answers to these calculations.

(a) 74×38.6

(b) 7.4×38.6

(c) $2856.4 \div 386$

4 The table shows the cost per minute, in pence, of phone calls to various parts of the world.

Country	Price per minute (pence)
Australia	3.5
Canada	4.2
China	3
France	3.5
Germany	3.5
Hungary	4.5
India	12.5
Ireland	3.5
Italy	3.5
Jamaica	7.5
Lithuania	9
New Zealand	4.2
Pakistan	16.7
Philippines	14.2
Poland	3.5
Russia	5
South Africa	6
Spain	3.5
Thailand	7.5
UK	1.2
United States	3.5
Zimbabwe	6.5

(a) Work out the cost, in pence, of making these calls.

(i)   24 minute call to Canada

(ii)   18 minute call to Pakistan

(iii)   43.2 minute call to Thailand

(b) Work out the length of a call, in minutes, to these places.

(i)   South Africa costing 327.6p

(ii)   Australia costing 84p

(iii)   UK costing 58.32p.

 5 Bob is laying three courses of 20-centimetre bricks.
The wall is 7 metres long and he takes 3 minutes to lay each brick.
How long will it take him to lay all these bricks?

 6 Annie buys 5.5 kg of potatoes at 76p per kilogram, 3 kg of carrots at 37p per kilogram and 4 kg of swede.
She gave a £10 note and received £3.03 change.
What was the cost of each kilogram of swede?

Chapter 2 — Angles, triangles and quadrilatrals

Exercise 2.1H

1 Calculate the angles marked with letters in these triangles.

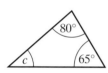

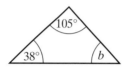

2 PQR is a triangle.
Angle QPR = 48° and angle PQR = 60°.
Sketch the triangle and find angle PRQ.

 3 ABCD is a rectangle.
P is halfway between B and C.

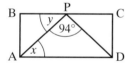

(a) What type of triangle is APD?
(b) Calculate angle x.
(c) What is the size of angle BAD?
(d) Calculate angle y.

 4 Harry says he has measured the angles of a triangle and they are 68°, 75° and 47°.
Is he correct?

 5 Martha measures the angles of a triangle and says they are 108°, 41° and 113°.
How can you tell, without doing any calculations, that she has made a mistake?

Exercise 2.2H

1 Look at these shapes.

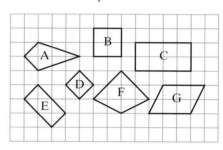

Copy and complete this table. Put a tick (✓) if the description fits. Put a cross (✗) if the description does not fit.

Shape	All sides equal	Opposite sides parallel	All angles right angles
A			
B			
C			
D			
E			
F			
G			

2 Name each of the shapes in question **1**.

3 (a) What is the name of a quadrilateral that has two pairs of equal sides which are next to each other?
(b) What other fact is true about this quadrilateral?

4 Sketch a trapezium.
Label clearly any equal sides and any parallel sides.

5 A quadrilateral has opposite sides equal and parallel.
Name all the different shapes this quadrilateral could be.

6 Which quadrilaterals have diagonals equal in length?

 7 A quadrilateral has angles of 110°, 70°, 110° and 70° as you go round the quadrilateral.
The sides are all the same length.
What special type of quadrilateral is this?
You may find it helpful to sketch the quadrilateral.

Chapter 3 | Fractions

Exercise 3.1H

1 Copy and complete each set of equivalent fractions.

(a) $\frac{3}{4} = \frac{\square}{8} = \frac{9}{\square} = \frac{30}{\square}$

(b) $\frac{4}{5} = \frac{\square}{10} = \frac{12}{\square} = \frac{36}{\square}$

(c) $\frac{1}{3} = \frac{\square}{6} = \frac{3}{\square} = \frac{\square}{12} = \frac{10}{\square} = \frac{\square}{300}$

(d) $\frac{1}{7} = \frac{2}{\square} = \frac{\square}{21} = \frac{4}{\square} = \frac{\square}{70} = \frac{100}{\square}$

2 Copy and complete each pair of equivalent fractions.

(a) $\frac{3}{5} = \frac{\square}{15}$

(b) $\frac{10}{18} = \frac{5}{\square}$

(c) $\frac{1}{2} = \frac{\square}{24}$

(d) $\frac{50}{70} = \frac{5}{\square}$

(e) $\frac{12}{22} = \frac{\square}{11}$

(f) $\frac{4}{7} = \frac{12}{\square}$

(g) $\frac{5}{11} = \frac{\square}{66}$

(h) $\frac{3}{27} = \frac{1}{\square}$

(i) $\frac{2}{9} = \frac{\square}{36}$

(j) $\frac{4}{13} = \frac{\square}{39}$

(k) $\frac{25}{45} = \frac{5}{\square}$

(l) $\frac{63}{70} = \frac{\square}{10}$

Exercise 3.2H

1 Express these fractions in their lowest terms.

(a) $\frac{10}{14}$

(b) $\frac{3}{18}$

(c) $\frac{16}{20}$

(d) $\frac{18}{30}$

(e) $\frac{21}{35}$

(f) $\frac{30}{50}$

(g) $\frac{24}{36}$

(h) $\frac{32}{56}$

(i) $\frac{4}{52}$

(j) $\frac{70}{140}$

(k) $\frac{20}{160}$

(l) $\frac{200}{1000}$

(m) $\frac{18}{54}$

(n) $\frac{42}{56}$

(o) $\frac{60}{72}$

(p) $\frac{75}{125}$

Exercise 3.3H

1 For each pair of fractions
- find the lowest common denominator.
- state which is the bigger fraction.

(a) $\frac{7}{8}$ or $\frac{3}{4}$

(b) $\frac{5}{9}$ or $\frac{7}{11}$

(c) $\frac{1}{6}$ or $\frac{3}{20}$

(d) $\frac{2}{3}$ or $\frac{4}{5}$

(e) $\frac{2}{7}$ or $\frac{3}{10}$

(f) $\frac{5}{8}$ or $\frac{2}{3}$

2 Katy and Natalie buy two identical pizzas.
Katy eats $\frac{2}{5}$ of her pizza.
Natalie eats $\frac{3}{8}$ of her pizza.
Who eats more pizza?

3 Write each of these sets of fractions in order, smallest first.

(a) $\frac{1}{2}$ $\frac{3}{10}$ $\frac{2}{5}$ $\frac{11}{20}$

(b) $\frac{2}{3}$ $\frac{3}{4}$ $\frac{5}{6}$ $\frac{7}{12}$

(c) $\frac{3}{5}$ $\frac{7}{10}$ $\frac{2}{3}$ $\frac{11}{15}$

Exercise 3.4

1 Find $\frac{1}{3}$ of these quantities.

(a) 30

(b) 36

(c) 69

(d) £90

(e) £15

2 Find $\frac{1}{5}$ of these quantities.

(a) 35

(b) 55

(c) 90

(d) £140

(e) 36 m

3 Find $\frac{3}{4}$ of these quantities.

(a) 36

(b) 60

(c) 100

(d) £68

(e) £180

4 Find $\frac{5}{6}$ of these quantities.

(a) 60

(b) 36

(c) 90

(d) 48 cm

(e) £150

5 Faheem receives £24 for his birthday.
He saves $\frac{1}{4}$ of it.
How much does he save?

6 Each week Abigail earns £20.
She spends $\frac{1}{5}$ of it on sweets and drinks.
She saves $\frac{3}{10}$ of it.
(a) How much does she spend on sweets and drinks?
(b) How much does she save?

7 A school's maths budget is £800.
$\frac{3}{5}$ is spent on text books.
How much is left?

8 There are 180 students in Year 9.
$\frac{4}{9}$ of them come to school by bus.
How many come by bus?

9 A school has 720 students.
$\frac{5}{8}$ of them have school dinners.
How many students have school dinners?

10 Show which is the larger, a $\frac{3}{8}$ share of £160 or a $\frac{4}{5}$ share of £80.

11 Jim left £1200 in his will.
He left $\frac{1}{4}$ to his son, $\frac{2}{5}$ to his wife and $\frac{7}{20}$ to his nephew.
How much does each person receive?

Exercise 3.5H

1 What fraction is
(a) 4 of 12? (b) 18 of 24?
(c) 9 of 15? (d) 14 of 21?
(e) 18 of 36? (f) £20 of £80?
(g) 7 cm of 56 cm? (h) 22 g of 77 g?

Write each fraction in its lowest terms.

Exercise 3.6H

For questions **1** to **5**, give your answers in their lowest terms.

1 Work out these.
(a) $\frac{3}{7} + \frac{2}{7}$ (b) $\frac{7}{15} + \frac{4}{15}$ (c) $\frac{8}{11} - \frac{3}{11}$
(d) $\frac{11}{17} - \frac{8}{17}$ (e) $\frac{7}{16} + \frac{3}{16}$ (f) $\frac{1}{9} + \frac{4}{9}$
(g) $\frac{7}{12} - \frac{5}{12}$ (h) $\frac{4}{11} + \frac{5}{11}$

2 Add these fractions.
(a) $\frac{2}{9} + \frac{4}{9}$ (b) $\frac{2}{5} + \frac{3}{10}$ (c) $\frac{1}{4} + \frac{1}{5}$
(d) $\frac{1}{3} + \frac{3}{10}$ (e) $\frac{3}{8} + \frac{1}{6}$

3 Subtract these fractions.
(a) $\frac{3}{8} - \frac{1}{8}$ (b) $\frac{5}{8} - \frac{1}{4}$ (c) $\frac{1}{3} - \frac{1}{8}$
(d) $\frac{5}{8} - \frac{1}{6}$ (e) $\frac{5}{8} - \frac{2}{5}$

4 Work out these.
(a) $\frac{2}{9} + \frac{1}{3}$ (b) $\frac{7}{12} + \frac{1}{4}$ (c) $\frac{3}{4} - \frac{1}{10}$
(d) $\frac{13}{16} - \frac{3}{8}$ (e) $\frac{5}{8} + \frac{1}{3}$ (f) $\frac{1}{6} + \frac{5}{8}$
(g) $\frac{7}{12} - \frac{1}{8}$ (h) $\frac{3}{20} + \frac{3}{4}$ (i) $\frac{4}{11} + \frac{3}{5}$
(j) $\frac{5}{12} + \frac{3}{10}$ (k) $\frac{7}{8} - \frac{1}{6}$ (l) $\frac{7}{15} - \frac{3}{20}$

5 Work out these.
(a) $\frac{2}{5} + \frac{1}{4} - \frac{1}{2}$
(b) $\frac{3}{8} + \frac{3}{4} - \frac{2}{3}$
(c) $\frac{1}{3} + \frac{1}{4} + \frac{1}{5}$

 6 Three friends shared the money they won on the lottery.
Ann received $\frac{2}{5}$, Malcolm received $\frac{1}{4}$ and Steve received the rest.
What fraction did Steve receive?

 7 Adam went on a three-day journey.
He drove $\frac{1}{3}$ of the distance on the first day and $\frac{3}{8}$ of the distance on the second day.
What fraction was left to drive on the third day?

Exercise 3.7H

1 Change each of these fractions to a decimal.
If necessary, give your answer to 3 decimal places.
(a) $\frac{1}{4}$ (b) $\frac{5}{8}$ (c) $\frac{11}{20}$
(d) $\frac{6}{10}$ (e) $\frac{7}{8}$ (f) $\frac{9}{12}$
(g) $\frac{5}{7}$ (h) $\frac{7}{9}$

Solving problems

Exercise 4.1H

1 How many hours and minutes are there between the following times?
   (a) 08:10 and 08:43
   (b) 15:00 and 19:00
   (c) 10:42 and 11:23
   (d) 13:48 and 15:22
   (e) 06:41 and 14:25
   (f) 22:10 on Thursday and 04:20 on Friday
   (g) $\frac{1}{4}$ to 4 and $\frac{1}{2}$ past 5
   (h) 20 minutes to 5 and 25 minutes past 8

2 Felicity arrived at the dentist at 12:40.
   She was at the dentist for 35 minutes.
   At what time did she leave?

3 Gary went to town on a bus.
   He arrived at the bus stop at 09:55 and waited 8 minutes for the bus.
   The bus took 23 minutes to get to town.
   At what time did Gary arrive in town?

4 (a) On weekdays a train leaves Sheffield for London at 13:27 and arrives in London at 15:50.
       How long does the journey take?
   (b) On Sundays a train leaves Sheffield for London at 12:51.
       It takes 3 hours and 13 minutes.
       At what time does the train arrive in London?

5 James has an appointment for 10:15.
   He arrives 16 minutes early.
   At what time does he arrive?

6 At what time do these TV programmes finish?
   (a) Starts at 08:15 and lasts 30 minutes.
   (b) Starts at 12:20 and lasts 50 minutes.
   (c) Starts at 18:25 and lasts 40 minutes.
   (d) Starts at 20:33 and lasts 1 hour 35 minutes.
   (e) Starts at 22:35 on Tuesday and lasts 2 hours 20 minutes.

7 Rupinder goes for lunch at 12:35 and returns at 13:47.
   For how long is she away?

8 Adam arrives 8 minutes early for a meeting due to start at 19:30.
   It starts 3 minutes late and finishes at 21:18.
   (a) At what time did Adam arrive?
   (b) How long was the meeting?

9 Rachel runs the 10 000 m race in 38 minutes 14 seconds.
   Caroline runs it in 39 minutes and 2 seconds.
   How much longer did it take Caroline?

10 Emily runs in a marathon.
   She checks the exact time when she crosses the start line. It is 10:41:05.
   She completes the race in a time of 4 hours 13 minutes and 6 seconds.
   What time is it when she finishes?

Exercise 4.2H

Use these approximate conversions for changing between imperial and metric units.

Length	Weight
8 km ≈ 5 miles	1 kg ≈ 2 pounds (lb)
1 m ≈ 40 inches	25 g ≈ 1 ounce (oz)
1 inch ≈ 2.5 cm	**Capacity**
1 foot (ft) ≈ 30 cm	4 litres ≈ 7 pints (pt)

1 Petra went on a 12 km hike.
How far was this in miles?

2 For a party Jasmine bought a 2 litre carton
of full cream milk and four 2 litre bottles of
semi-skimmed milk.
How many pints of milk is that altogether?

3 Simon's rucksack has a mass of 32 pounds.
How much is this in kilograms?

4 Fitz bought four lengths of wood, each 8 foot
long.
What was the total length in metres and
centimetres?

5 At the swimming baths there is a 5-metre diving
board.
How high is this in feet and inches?

6 The distance from Sheffield to Cambridge is
185 miles via the M1, but only 148 miles if you
cut across to the A1.
How much further, in kilometres, is the journey
via the M1?

7 My car holds 10 gallons of petrol.
1 gallon is 8 pints.
How many litres does it hold?
Give your answer to the nearest litre.

8 On a flight passengers are each allowed to have
luggage weighing up to 20 kg.
A more accurate conversion is 1 kg = 2.2 lbs.
Use this to find how much they are each allowed
in pounds.

9 Colin drives 320 miles to a meeting.
Ruth drives 504 km to the same meeting.
Who has driven further and by how much?

Exercise 4.3H

1 On holiday in France, Jason made this table of
approximate conversions between kilometres per
hour (km/h) and miles per hour (mph).

km/h	40	50	60	70	80	90	100	110	120
mph	25	30	40	45	50	55	60	65	75

A more accurate conversion is
mph = km/h × 0.625.
(a) Jason saw a speed limit sign of 90 km/h.
What is the speed limit in mph
(i) according to his table?
(ii) using the more accurate conversion?
(b) Some of the conversions in the table are
exactly the same as using the accurate
conversion. List them.

2 Here is part of a Sunday timetable for trains from
Sheffield and Chesterfield to London St Pancras.

Sheffield	Chesterfield	London St Pancras
0853	0906	1219
0951	1007	1318
1102	1113	1418
1251	1303	1604
1353	1407	1706
1443	1456	1755
1601	1616	1914
1830	1843	2141

(a) What time does the 1102 from Sheffield
arrive at London St Pancras?
(b) What time does the train arriving in London
at 1755 leave Chesterfield?
(c) Which of the trains in the list completes the
journey from Sheffield to London in the
shortest time?

3 The table shows the cost per minute of phone calls to various parts of the world.

Country	Price per minute	Country	Price per minute
Australia	3.5	New Zealand	4.16
Canada	4.16	Pakistan	16.66
China	3	Philippines	14.18
France	3.5	Poland	3.5
Germany	3.5	Russia	5
Hungary	4.5	South Africa	6
India	12.5	Spain	3.5
Ireland	3.5	Thailand	7.5
Italy	3.5	UK	3.5
Jamaica	7.5	United States	3.5
Lithuania	9	Zimbabwe	6.5

(a) How much does a 10-minute call to Spain cost?

(b) Which other countries cost the same to phone as France?

(c) Selina wants to phone her cousin in South Africa.
She has only a pound to spend.
What is the greatest number of complete minutes she can phone for?

4 Look at the distance chart.

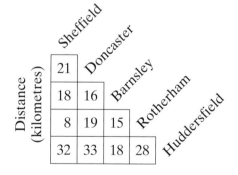

(a) How far is it from Doncaster to Huddersfield?

(b) Which two towns are the closest?

(c) How much further is it to go from Sheffield to Huddersfield via Barnsley rather than direct?

5 This table shows the cost per person for going to Hotel Regina in Cyprus.

Basic costs per person in pounds				
Number of nights	**7 nights**		**14 nights**	
	Adult	**Child**	**Adult**	**Child**
1–30 April	350	180	450	210
1–22 May	365	210	475	240
23–31 May	480	250	560	280
1–30 June	385	220	415	250
1–19 July	425	240	525	275
20 July–25 Aug	515	285	615	315
16 Aug–30 Sept	385	220	485	245
1–31 Oct	350	180	420	195

Departure date between

(a) Find the cost for an adult departing for 14 nights on June 5th.
(b) Find the cost of a family of two adults and three children departing for seven nights on 18 July.
(c) (i) How much extra would it cost for the family in part **(b)** to go one week later?
  (ii) Why do you think it costs so much more?

Angles

Exercise 5.1H

Work out the size of the unknown angle in each of these diagrams.

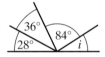

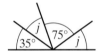

Exercise 5.2H

Work out the size of the unknown angle in each of these diagrams.

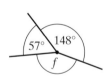

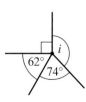

Exercise 5.3H

For each question

- work out the size of each unknown angle.
- give a reason for each answer.

1

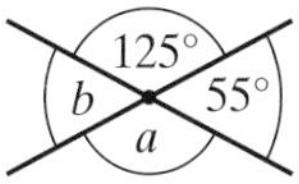

2

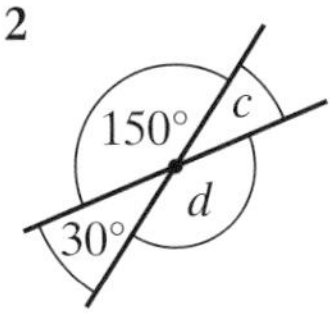

3

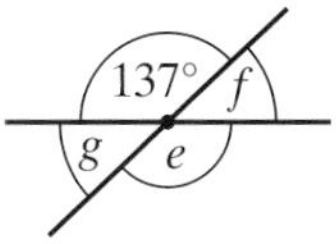

4

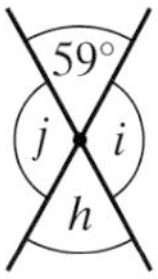

5

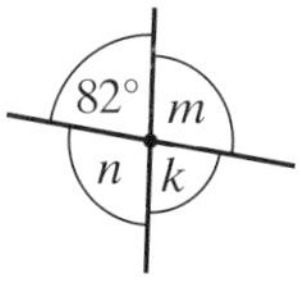

6

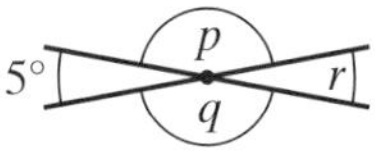

7

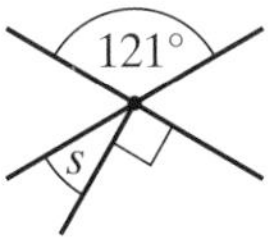

8

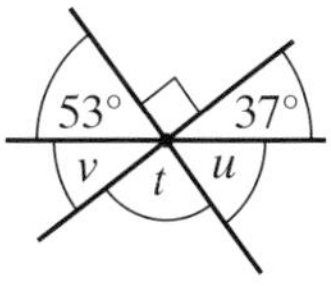

9

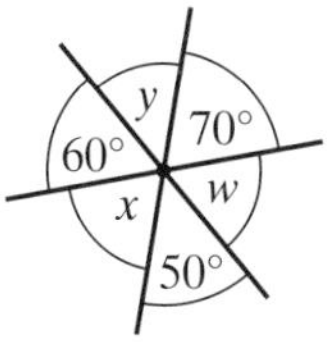

10

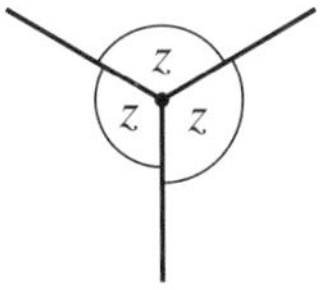

Exercise 5.4H

Find the size of the lettered angles.
Give a reason for each answer.

1

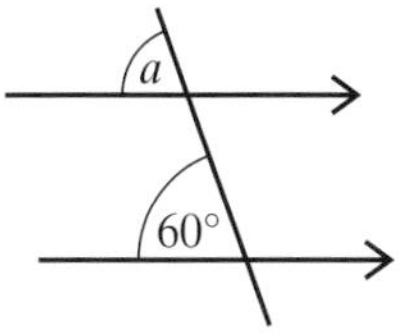

2

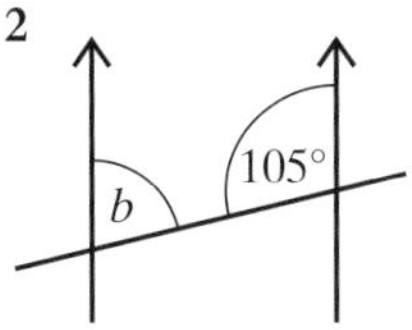

3

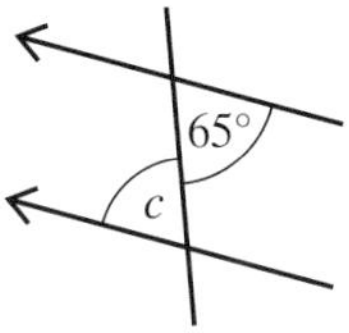

4

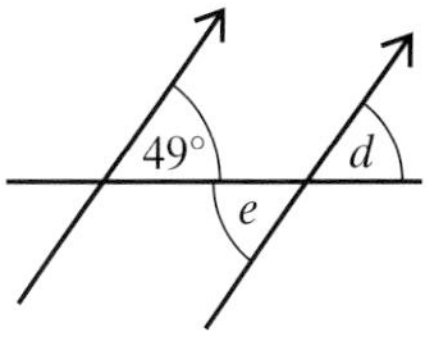

5

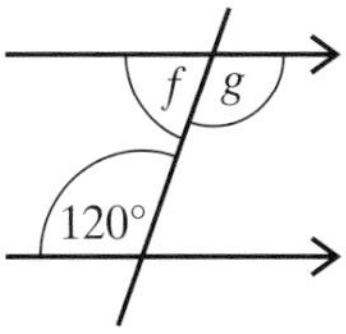

6

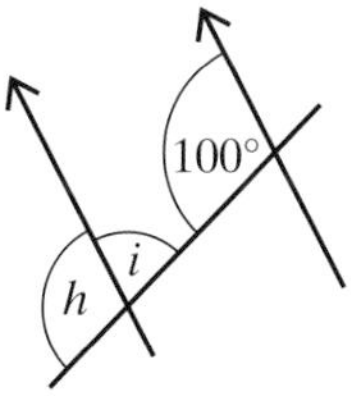

7

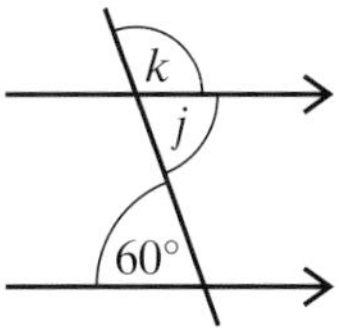

8

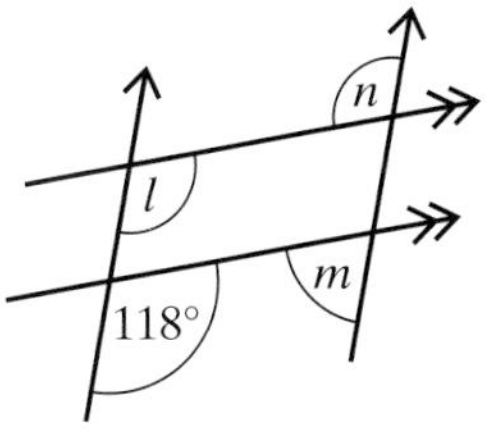

9

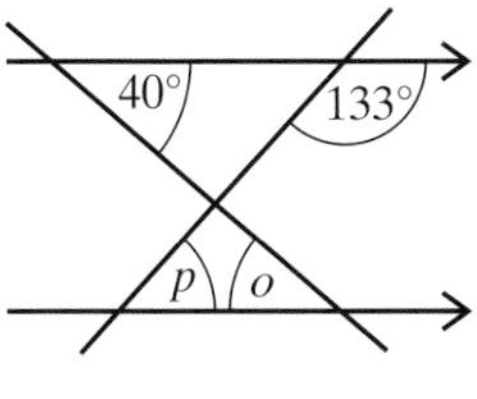

10

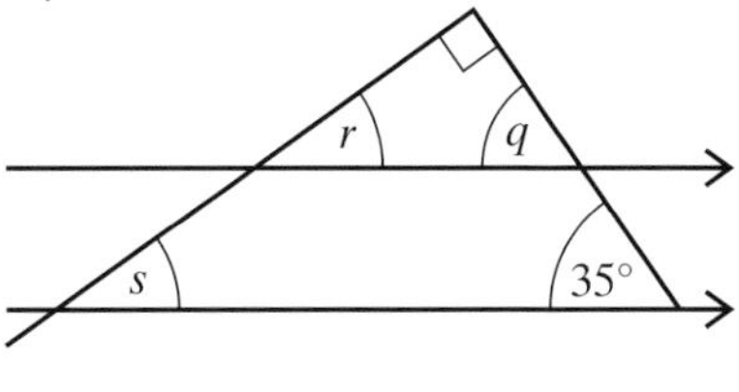

Exercise 6.1H

1 Change each of these improper fractions to a mixed number.

(a) $\frac{7}{2}$ (b) $\frac{10}{3}$ (c) $\frac{17}{8}$

(d) $\frac{14}{9}$ (e) $\frac{19}{5}$ (f) $\frac{10}{7}$

(g) $\frac{12}{7}$ (h) $\frac{11}{2}$ (i) $\frac{13}{4}$

(j) $\frac{11}{3}$

2 Change each of these mixed numbers to an improper fraction.

(a) $1\frac{2}{7}$ (b) $1\frac{5}{8}$ (c) $7\frac{1}{2}$

(d) $2\frac{3}{4}$ (e) $3\frac{2}{5}$ (f) $2\frac{4}{5}$

(g) $3\frac{2}{9}$ (h) $4\frac{5}{6}$ (i) $5\frac{1}{8}$

(j) $4\frac{1}{3}$

3 Write these as simply as possible (as fractions or mixed numbers in their lowest terms).

(a) $\frac{18}{10}$ (b) $\frac{15}{9}$ (c) $\frac{28}{24}$

(d) $\frac{7}{28}$ (e) $\frac{9}{6}$ (f) $\frac{10}{25}$

(g) $\frac{25}{15}$ (h) $\frac{24}{18}$ (i) $\frac{65}{55}$

(j) $\frac{60}{45}$

Exercise 6.2H

Work out these.

Where necessary, write your answers as proper fractions or mixed numbers in their lowest terms.

1 $\frac{1}{4} \times 32$ **2** $\frac{1}{5} \times 55$

3 $\frac{3}{4} \times 60$ **4** $\frac{4}{5} \times 100$

5 $\frac{1}{2} \times 6$ **6** $\frac{1}{4} \times 10$

7 $\frac{2}{3} \times 6$ **8** $\frac{5}{8} \times 16$

9 $\frac{1}{9} \times 5$ **10** $\frac{1}{12} \times 4$

11 $\frac{1}{4} \times \frac{3}{5}$ **12** $\frac{1}{5} \times \frac{1}{4}$

13 $\frac{3}{4} \times \frac{2}{3}$ **14** $\frac{4}{5} \times \frac{1}{2}$

15 $\frac{3}{4} \times \frac{8}{9}$ **16** $\frac{5}{6} \times \frac{3}{5}$

17 $\frac{2}{3} \times \frac{5}{7}$ **18** $\frac{1}{8} \times \frac{5}{6}$

19 $\frac{7}{12} \times \frac{3}{8}$ **20** $\frac{7}{30} \times \frac{10}{21}$

21 A box of chocolates has 20 chocolates in it. Ben has eaten $\frac{3}{5}$ of them. How many chocolates has Ben eaten?

22 After a birthday party, $\frac{3}{8}$ of the cake is left over. When Dad gets home he eats $\frac{1}{3}$ of what is left.

What fraction of the whole cake does Dad eat? Give your answer in its lowest terms.

Exercise 6.3H

Work out these.
Where necessary, write your answers as proper
fractions or mixed numbers in their lowest terms.

1 $\frac{2}{9} \div 4$　　　　　　　2 $9 \div \frac{3}{8}$

3 $\frac{1}{4} \div \frac{3}{5}$　　　　　　　4 $\frac{4}{5} \div \frac{3}{5}$

5 $5 \div \frac{1}{6}$　　　　　　　6 $\frac{4}{9} \div 3$

7 $\frac{3}{4} \div \frac{4}{5}$　　　　　　　8 $\frac{5}{9} \div \frac{5}{6}$

9 $\frac{8}{19} \div \frac{4}{9}$　　　　　10 $\frac{3}{5} \div \frac{13}{15}$

11 $\frac{2}{5} \div \frac{8}{15}$　　　　　12 $\frac{1}{4} \div \frac{2}{5}$

13 $\frac{4}{5} \div \frac{14}{15}$　　　　　14 $\frac{9}{21} \div \frac{3}{7}$

15 $\frac{3}{8} \div \frac{1}{3}$　　　　　16 $\frac{4}{9} \div \frac{1}{5}$

17 $\frac{6}{7} \div \frac{1}{8}$　　　　　18 $\frac{7}{15} \div \frac{2}{3}$

19 $\frac{9}{16} \div \frac{7}{12}$　　　　20 $\frac{7}{10} \div \frac{5}{12}$

21 Dimba the dog eats $\frac{3}{4}$ of a tin of dog meat
every day.
How many tins should his owner buy to feed
Dimba for 7 days?

22 In a competition, the Blue team got 360 points.
This was $\frac{4}{9}$ of the total number of points awarded.
The Red team got $\frac{2}{5}$ of the points.
How many points did the Red team get?
Which team did better?

Exercise 6.4H

Work out these.

1 (a) $1\frac{1}{4} \times \frac{3}{5}$　　　　　(b) $2\frac{1}{2} \times 1\frac{3}{5}$
　　(c) $4\frac{1}{4} \times 2\frac{2}{5}$　　　　(d) $3\frac{2}{3} \times 2\frac{1}{4}$
　　(e) $3\frac{1}{6} \times \frac{3}{5}$　　　　　(f) $4\frac{1}{5} \times 1\frac{3}{14}$

2 (a) $3\frac{1}{4} \div 1\frac{1}{8}$　　　　　(b) $4\frac{1}{6} \div 1\frac{3}{7}$
　　(c) $6\frac{2}{5} \div \frac{8}{15}$　　　　　(d) $3\frac{1}{6} \div 1\frac{1}{9}$
　　(e) $2\frac{3}{4} \div 5\frac{1}{2}$　　　　　(f) $2\frac{2}{7} \div 3\frac{2}{3}$

3 (a) $4\frac{1}{2} \times 2\frac{1}{6}$　　　　　(b) $5\frac{1}{4} \times 3\frac{1}{7}$
　　(c) $5\frac{3}{4} \div \frac{5}{8}$　　　　　(d) $8 \div 3\frac{1}{3}$
　　(e) $5\frac{1}{7} \times 1\frac{5}{9}$　　　　(f) $1\frac{3}{5} \div 4\frac{1}{10}$
　　(g) $3\frac{5}{9} \times 2\frac{5}{8}$　　　　(h) $4\frac{1}{6} \div 1\frac{2}{9}$
　　(i) $4\frac{2}{7} \times 1\frac{5}{16}$　　　(j) $3\frac{3}{4} \div 1\frac{5}{16}$
　　(k) $4\frac{1}{7} \div 1\frac{3}{14}$　　　(l) $5\frac{1}{3} \times 2\frac{5}{8}$

4 (a) $1\frac{3}{4} \times 2\frac{1}{2} \times \frac{8}{15}$　　(b) $1\frac{1}{3} \times 3\frac{2}{5} \div 1\frac{5}{6}$
　　(c) $4\frac{2}{7} \times 2\frac{4}{5} \div 2\frac{1}{4}$　　(d) $4\frac{1}{2} \times 2\frac{1}{3} \times \frac{5}{14}$
　　(e) $5\frac{1}{4} \times 2\frac{2}{3} \div 2\frac{2}{9}$　　(f) $5\frac{1}{3} \times 2\frac{3}{4} \div 3\frac{3}{10}$

Exercise 6.5H

1 Add each of these.
   Write your answers as simply as possible.

   (a) $\frac{3}{10} + 1\frac{2}{5}$ **(b)** $1\frac{1}{4} + \frac{3}{5}$ **(c)** $2\frac{1}{5} + 1\frac{1}{3}$

   (d) $4\frac{1}{2} + 2\frac{3}{5}$ **(e)** $1\frac{5}{6} + \frac{2}{5}$

2 Subtract each of these.
   Write your answers as simply as possible.

   (a) $2\frac{3}{10} - \frac{1}{10}$ **(b)** $3\frac{5}{6} - 1\frac{3}{8}$ **(c)** $4\frac{1}{2} - 2\frac{1}{8}$

   (d) $3\frac{1}{2} - \frac{3}{5}$ **(e)** $6\frac{1}{10} - 4\frac{2}{5}$

3 Work out these.
   Write your answers as simply as possible.

   (a) $5\frac{2}{3} + 3\frac{3}{4}$ **(b)** $3\frac{5}{8} - 2\frac{1}{6}$ **(c)** $6\frac{7}{12} - 3\frac{7}{8}$

   (d) $2\frac{9}{16} + 4\frac{7}{8}$ **(e)** $8\frac{2}{9} - 2\frac{5}{6}$ **(f)** $7\frac{2}{5} + 2\frac{4}{7}$

   (g) $5\frac{6}{7} + 2\frac{1}{2}$ **(h)** $3\frac{3}{8} - 2\frac{7}{10}$

4 Work out these.
   Write your answers as simply as possible.

   (a) $5\frac{1}{2} + 3\frac{5}{9} - 4\frac{2}{3}$ **(b)** $6\frac{2}{5} - 2\frac{7}{10} + 3\frac{1}{2}$

5 A shopkeeper had a roll of cloth $16\frac{1}{2}$ metres long.
   Sasha bought a piece $2\frac{3}{4}$ metres long.
   How much was left?

6 A shopkeeper sells wire netting by the metre.
   He buys it in 10-metre rolls.

   Two customers buy $2\frac{2}{5}$ and $3\frac{1}{4}$ metres respectively.

   A third customer wants $4\frac{1}{2}$ metres.

   Will the shopkeeper need to start a new roll?
   Show your working.

Exercise 7.1H

1 Name these parts of a circle.

(a)   (b)   (c)

2 Name these polygons.

(a)   (b)   (c)

3 Draw any hexagon.
Draw diagonals across the hexagon from each vertex (corner) of the hexagon to another vertex. How many diagonals can you draw in the hexagon altogether?

4 Draw any seven-sided polygon.
Draw diagonals across the polygon from each vertex of the polygon to another vertex. How many diagonals can you draw in the polygon altogether?

5 A regular hexagon is constructed in a circle. How many degrees are measured at the centre to draw each radius required?

6 A 20-sided regular polygon is constructed in a circle.
How many degrees are measured at the centre to draw each radius required?

7 Draw a circle of radius 5 cm and use it to construct a regular pentagon.
Measure the length of a side of your pentagon.

8 Draw a circle of radius 6 cm and use it to construct a regular nine-sided polygon.
Measure the length of a side of your polygon.

Exercise 7.2H

Find the size of the lettered angles.
Give a reason for each answer.

1

2

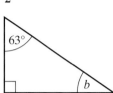

3

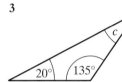

4

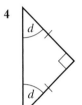

5

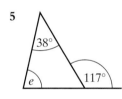

6

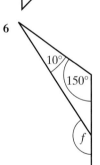

7

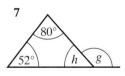

8

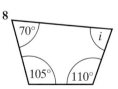

9

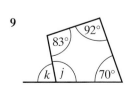

10

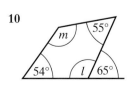

Exercise 7.3H

1 A polygon has nine sides.
Work out the sum of the interior angles of this polygon.

2 A polygon has 13 sides.
Work out the sum of the interior angles of this polygon.

3 Four of the exterior angles of a hexagon are 93°, 50°, 37° and 85°.
The other two angles are equal.

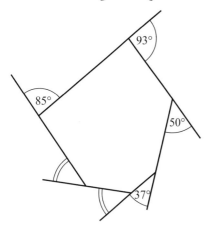

(a) Work out the size of these equal exterior angles.
(b) Work out the size of the interior angles of the hexagon.

4 Four of the interior angles of a pentagon are 170°, 80°, 157°, and 75°.

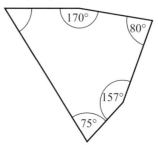

(a) Work out the size of the other interior angle.
(b) Work out the exterior angles of the pentagon.

5 A regular polygon has 18 sides.
Find the size of the exterior and interior angles of this polygon.

6 A regular polygon has 24 sides.
Find the size of the exterior and interior angles of this polygon.

7 A regular polygon has an exterior angle of 12°.
Work out the number of sides that the polygon has.

8 A regular polygon has an interior angle of 172°.
Work out the number of sides that the polygon has.

Powers and indices

Exercise 8.1H

1 Write down the square of each of these numbers.
(a) 3   (b) 4   (c) 7
(d) 9   (e) 12

2 Write down the positive square root of each of these numbers.
(a) 4   (b) 25   (c) 36
(d) 64   (e) 121

3 Write down the positive and negative square roots of each of these numbers.
(a) 144   (b) 1   (c) 49
(d) 9   (e) 81

4 Work out these.
(a) 6^2 (b) 1^2 (c) 10^2
(d) 5^2 (e) 11^2

5 Work out these.
(a) $\sqrt{49}$ (b) $\sqrt{9}$ (c) $\sqrt{81}$
(d) $\sqrt{100}$ (e) $\sqrt{1}$

6 Work out these.
(a) 40^2 (b) 70^2

7 Work out these.
(a) $8^2 - 6^2$ (b) $4^2 + 5^2$ (c) $7^2 - 3^2$
(d) $6^2 - 3^2$ (e) $7^2 + 4^2$ (f) $5^2 - 2^2$

8 Write down the cube of each of these numbers.
(a) 2   (b) 5   (c) 10
(d) 3   (e) $\sqrt[3]{15}$

9 Write down the cube root of each of these numbers.
(a) 64   (b) 1   (c) 8
(d) 1000   (e) 30^3

10 A square has an area of $400\,\text{cm}^2$.
What is the length of its sides?

Exercise 8.2H

1 Write each of these using index notation.
(a) $4 \times 4 \times 4$
(b) $7 \times 7 \times 7 \times 7$
(c) $3 \times 3 \times 3 \times 3 \times 3 \times 3 \times 3 \times 3 \times 3$
(d) $9 \times 9 \times 9 \times 9 \times 9 \times 9$
(e) $2 \times 2 \times 2 \times 2 \times 2 \times 2 \times 2$

2 Find the value of each of these.
(a) 7^3 (b) 2^3 (c) 3^3
(d) 1^3 (e) 5^3 (f) 2^4
(g) 1^6 (h) 4^5 (i) 7^4
(j) 2^8

Exercise 8.3H

1 Work out these, giving your answers in index form.
(a) $2^2 \times 2^4$ (b) $3^6 \times 3^2$ (c) $4^2 \times 4^3$
(d) $5^6 \times 5$

2 Work out these, giving the answers in index form.
(a) $5^5 \div 5^2$ (b) $7^8 \div 7^2$ (c) $2^6 \div 2^4$
(d) $3^7 \div 3^3$

3 Work out these, giving your answers in index form.
(a) $5^5 \times 5^3 \div 5^2$ (b) $10^4 \times 10^6 \div 10^5$
(c) $8^3 \times 8^3 \div 8^4$ (d) $3^5 \times 3 \div 3^3$

4 Work out these, giving your answers in index form.
(a) $\dfrac{2^5 \times 2^4}{2^3}$ (b) $\dfrac{3^7}{3^5 \times 3^2}$
(c) $\dfrac{5^5 \times 5^4}{5^2 \times 5^3}$ (d) $\dfrac{7^5 \times 7^2}{7^2 \times 7^4}$

Exercise 9.1H

1 Put these numbers in order of size, smallest first.
   (a) 0.29,   0.902,   0.249,   0.9402,   0.021
   (b) 0.803,   0.083,   0.0083,   0.003 08, 0.38
   (c) 0.092,   0.409,   0.429,   0.0942,   0.9
   (d) 83.1,   4270,   0.92,   6.347,   762.53
   (e) 3.01,   0.103,   13.01,   1.003,   3100
   (f) 17.24,   70.14,   1472,   4.017,   0.741

2 Put these numbers in order of size, smallest first.

   (a) 0.33, $\frac{31}{100}$, $\frac{29}{100}$,   0.3,   0.32

   (b) 0.55, $\frac{57}{100}$, $\frac{56}{100}$,   0.5,   0.54

   (c) $3\frac{71}{100}$,   3.73, $3\frac{7}{10}$,   3.78,   3.8

   (d) $8\frac{43}{100}$, $8\frac{4}{100}$,   8.4,   8.47,   8.39

   (e) 0.25, $\frac{3}{10}$, $\frac{28}{100}$,   0.35,   0.27

   (f) 0.46, $\frac{64}{100}$, $\frac{56}{100}$,   0.45,   0.65

Exercise 9.2H

1 Write these decimals as fractions.
   (a) 0.7   (b) 0.83   (c) 0.25
   (d) 0.507   (e) 0.09   (f) 0.4
   (g) 3.01   (h) 0.013

2 Convert each of these decimals to a fraction in its
   lowest terms.
   (a) 0.23   (b) 0.6   (c) 0.95
   (d) 0.008   (e) 0.04   (f) 0.175

Exercise 9.3H

For this exercise use the facts below to help you.

$0.\dot{1} = \frac{1}{9}$ $0.\dot{0}\dot{1} = \frac{1}{99}$ $0.\dot{0}0\dot{1} = \frac{1}{999}$

1 Copy and complete these sentences.

   (a) $\frac{2}{9}$ is .... $\times \frac{1}{9}$ so it equals .... $\times 0.\dot{1} =$ ....

   (b) $\frac{8}{99}$ is .... $\times \frac{1}{99}$ so it equals .... $\times 0.\dot{0}\dot{1} =$ ....

   (c) $\frac{207}{999}$ is .... $\times \frac{1}{999}$ so it equals .... $\times = 0.\dot{0}0\dot{1}$

   = ...

2 Copy and complete these sentences.

   (a) $0.\dot{4}$ is .... $\times 0.\dot{1}$ so it equals .... $\times \frac{1}{9} =$ ....

   (b) $0.\dot{3}\dot{7}$ is .... $\times 0.\dot{0}\dot{1}$ so it equals .... $\times \frac{1}{99}$

   = ...

   (c) $0.\dot{5}7\dot{1}$ is .... $\times 0.\dot{0}0\dot{1}$ so it equals .... $\times \frac{1}{999}$

   = ...

3 Use the facts at the beginning of the exercise to
   change these fractions into decimals.
   (a) $\frac{5}{9}$ (b) $\frac{46}{99}$ (c) $\frac{7}{99}$
   (d) $\frac{385}{999}$ (e) $\frac{106}{999}$ (f) $\frac{8}{999}$

4 Use the facts at the beginning of the exercise to
   change these recurring decimals to fractions.
   Simplify them where possible.
   (a) $0.\dot{7}$ (b) $0.\dot{3}$ (c) $0.\dot{1}\dot{7}$
   (d) $0.\dot{4}\dot{5}$ (e) $0.\dot{7}4\dot{2}$ (f) $0.\dot{1}8\dot{6}$

 5 Divide the fractions below into two groups, one
   group of fractions that will terminate and the
   other that will recur when changed into
   decimals.

   $\frac{1}{80}$, $\frac{1}{54}$, $\frac{2}{39}$, $\frac{5}{16}$, $\frac{11}{27}$, $\frac{25}{72}$, $\frac{49}{88}$, $\frac{12}{25}$, $\frac{67}{160}$, $\frac{5}{12}$

Real-life graphs

Exercise 10.1H

1 This graph converts between feet and centimetres.
Use the graph to find how many
  (a) centimetres is 6 feet.
  (b) feet is 250 cm.
  (c) centimetres is 1 foot.

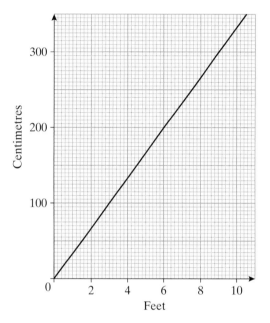

2 This graph converts between euros (€) and dollars ($).
Use the graph to find how many
  (a) dollars is €50.
  (b) euros is $25.
  (c) dollars is €1.

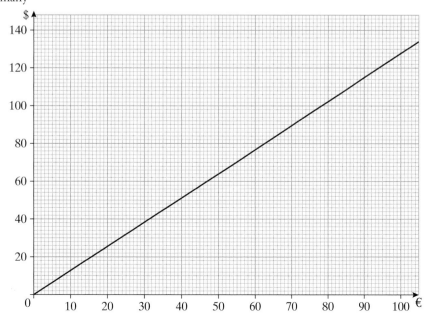

3 The area of land used to be measured in acres.
It is now measured in hectares (ha).
One hectare is $10\,000$ m^2.
The graph converts between acres and hectares.

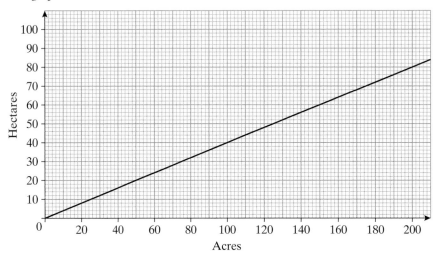

Use the graph to find how many
(a) hectares is 100 acres.
(b) acres is 45 ha
(c) acres is 1 ha.

4 This graph converts between centimetres
and inches.
Use the graph to find
(a) the number of inches equal to
   (i) 100 cm.
   (ii) 168 cm.
(b) the number of centimetres equal to
   (i) 50 inches.
   (ii) 21 inches.
(c) How many centimetres are equal
   to 200 inches?

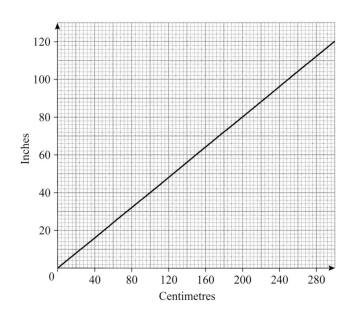

5 (a) On a piece of graph paper, mark axes horizontally for pounds (£) up to £100, using 1 cm to £10, and vertically for dollars ($) up to $160, using 1 cm to $20.

(b) Plot the point (100, 142) and join it to (0, 0) with a straight line.

(c) Find the number of dollars equal to
(i) £50. **(ii)** £85.

(d) Find the number of pounds equal to
(i) $100. **(ii)** $53.

(e) How many dollars are equal to £1?

Exercise 10.2H

1 Jack is walking from Church Stretton to Ratlinghope across the Long Mynd, a hill in Shropshire.
His journey is shown on the graph.

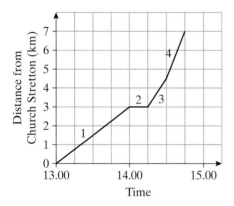

Describe each of the four stages of Jack's journey.

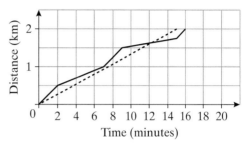

2 Jasbir and Robert have a race over 2 km. Robert runs, then walks and then runs again. Jasbir jogs at a steady speed. The graph shows what happened.

(a) Whose graph is the broken line?
(b) Who won the race?
(c) Describe what happened when the lines cross.

3 The graph shows two trains travelling between Leicester and St Pancras. The distance between the stations is 100 miles.

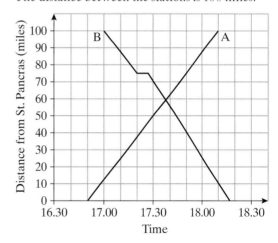

(a) When did train B leave Leicester?
(b) Which train stopped? How far from St Pancras was it when it stopped?
(c) Find the time and distance from St Pancras when the trains passed each other.

4 This graph shows Victoria's walk to her friend's house and back home.

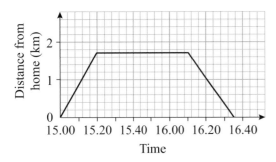

   (a) How far is it to Victoria's friend's house?
   (b) How long did she stay at her friend's house?
   (c) At what time did she get back home?
   (d) On which journey did she walk faster, going or coming back?

5 Jenny and George both travelled on the same road from Guildford to Leicester.
This graph shows both their journeys.

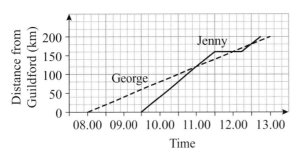

   (a) How much later than George did Jenny leave?
   (b) For how long did Jenny stop on the way?
   (c) What was George's average speed?
   (d) At what time did George and Jenny first pass each other?
   (e) How far were they from Guildford when they last passed each other?

6 (a) Draw a travel graph to show this cycle ride.

Use a scale of 1 cm for 30 minutes and 1 cm for 5 km.

Robert left home at 9 a.m. and cycled 12 km in the next 45 minutes.

He then stopped for 15 minutes.

He continued his journey and went 10 km in the next 45 minutes.

After a half hour's break he cycled directly home, arriving there at 1 p.m.

(b) Use your graph to answer these questions.

(i) How long did it take Robert to cycle back home?

(ii) How far did he cycle altogether?

(iii) What was his speed on the first section of his journey?

Chapter 11 Reflection

Exercise 11.1H

1 Copy these shapes.
On each shape, draw the lines of symmetry.

(a)   **(b)**   **(c)**

2 How many lines of symmetry do these arrows have?

(a)   **(b)**   **(c)**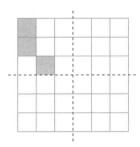

3 Draw a rectangle.
A rectangle has two lines of symmetry.
Draw both lines of symmetry on your rectangle.

4 This triangle has three lines of symmetry.
What type of triangle is it?

5 Copy this grid.
Shade more squares so that the diagonal broken line is the line of symmetry.

6 Copy this grid.
Shade more squares so that the grid has two lines of symmetry.

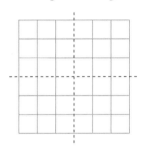

7 Make a pattern with two lines of symmetry.
Shade squares on a grid like this.

8 Copy these diagrams.
Complete the diagrams so that the broken lines are lines of symmetry.

(a)   **(b)**

(c)

9 Copy each diagram.
Complete the diagrams so that the dotted lines are lines of symmetry.
Each one spells a word or gives a number.

(a) COOK.   (b) BIKE.   (c) BID.   (d) 1308.

Exercise 11.2H

1 Copy each shape on to squared paper and reflect it in the mirror line shown.

(a)

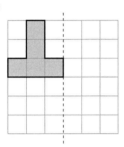

(b)

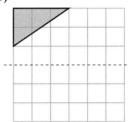

(c)

(d)

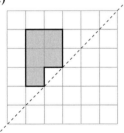

(e)

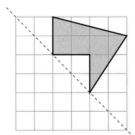

2 Copy each pair of shapes on to squared paper and draw the mirror line for the reflection.

(a)

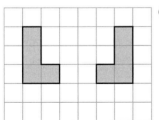

(b)

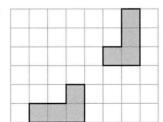

(c)

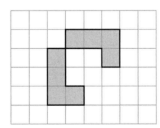

3 Copy the diagram.
Reflect flag A in the mirror line.
Label the image B.
Reflect flag A in the y-axis.
Label the image C.

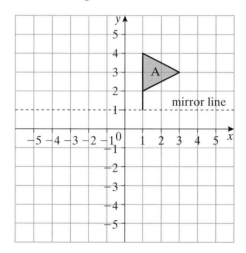

4 Draw x- and y-axes from -5 to 5.
Plot the points (2, 1), (5, 1), (5, 3) and (3, 5).
Join the points to form a trapezium.
Label it A.
Reflect shape A in the y-axis.
Label the image B.
Reflect shape A in the x-axis.
Label the image C.

5 Draw x- and y-axes from -2 to 4.
(a) Draw a triangle with vertices at (1, 1), (1, 3)
and (0, 3).
Label it A.
(b) Reflect triangle A in the line $x = 2$.
Label it B.
(c) Reflect triangle A in the line $y = x$.
Label it C.
(d) Reflect triangle A in the line $y = 2$.
Label it D.

6 Draw x- and y-axes from -3 to 3.
(a) Draw a triangle with vertices at $(-1, 1)$,
$(-1, 3)$ and $(-2, 3)$.
Label it A.
(b) Reflect triangle A in the line $x = \frac{1}{2}$.
Label it B.
(c) Reflect triangle A in the line $y = x$.
Label it C.
(d) Reflect triangle A in the line $y = -x$.
Label it D.

7 Describe fully the transformation that maps
(a) shape A on to shape B.
(b) shape A on to shape C.
(c) shape B on to shape D.

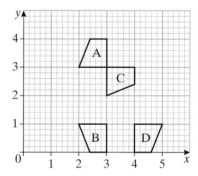

8 Describe fully the transformation that maps
(a) triangle A on to triangle B.
(b) triangle A on to triangle C.
(c) triangle E on to triangle F.

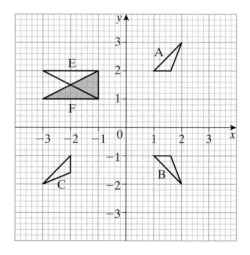

Chapter 12 Percentages

Exercise 12.1H

1. Change these percentages to fractions.
   Write your answers in their lowest terms.
   (a) 45%   (b) 85%   (c) 4%
   (d) 130%

2. Change these percentages to decimals.
   (a) 18%   (b) 26%   (c) 92%
   (d) 8%   (e) 42%   (f) 3%
   (g) 12%   (h) 1%   (i) 170%
   (j) 350%   (k) 5%   (l) 14.5%

3. Change these decimals to percentages.
   (a) 0.73   (b) 0.28   (c) 0.06
   (d) 0.235 (e) 1.68

4. Change these fractions to percentages.
   (a) $\frac{3}{10}$ (b) $\frac{4}{5}$ (c) $\frac{13}{20}$
   (d) $\frac{7}{25}$ (e) $\frac{18}{40}$

5. Put these numbers in order, smallest first.
   $\frac{7}{10}$ $\frac{17}{25}$ 72%   0.84 $\frac{2}{3}$

6. Put these numbers in order, smallest first.
   $\frac{1}{3}$ $\frac{7}{25}$ 30%   0.29   16.9%

7. At Tony's school, $\frac{8}{25}$ of the students support
   Manchester United.
   27% support Manchester City.
   Which team has the greater support?

8. In a test, Aftab got $\frac{13}{20}$ of the questions correct.
   Jamil got 63% correct.
   Who answered more of the questions correctly?

9. At Seahome it rains on $\frac{3}{20}$ of the days.
   At Welton it rains on $\frac{1}{6}$ of the days.
   Which place had rain on more days?

10. At Jim's school the football teams have had a
    successful season.
    The Under 14s won $\frac{5}{8}$ of their games.
    The Under 15s won 63% of their games.
    The Under 16s won 0.62 of their games.
    List the teams in order, most successful first.

Exercise 12.2H

Do not use your calculator for questions **1** to **4**.

1. (a) Find 25% of £84.
   (b) Find 60% of 35 kg.
   (c) Find 15% of £70.

2. James borrowed £80 000 to buy a house and paid
   7% interest in the first year.
   Calculate the interest.

3. 30% of the students in Year 7 came from one
   primary school.
   There are 220 students in Year 7.
   How many came from that primary school?

4. 15% of Mr and Mrs Thompson's Council Tax is
   used to pay for police, fire service and civil defence.
   Mr and Mrs Thompson's Council tax bill is £1200.
   How much goes to pay for police, fire service
   and civil defence?

You may use your calculator for questions **5** to **8**.

5. (a) Find 19% of £36.
   (b) Find 37% of 240 metres.
   (c) Find 108% of £64.

6. 92% of the seats at a concert were sold after
   1 week.
   There were 32 000 seats available.
   How many were sold in the first week?

7. Sarah pays 6% of her earnings into a pension fund.
   She earns £1450 per month.
   How much does she pay into the pension fund
   each month?

8. Michael changes £350 into euros.
   The bank charges him 2.5% commission.
   How much is the commission?

Exercise 12.3H

Do not use your calculator for questions **1** to **6**.

1 Increase £700 by these percentages.
   (a) 40%   **(b)** 35%   **(c)** 70%   **(d)** 21%

2 Decrease £520 by these percentages.
   (a) 70%   **(b)** 25%   **(c)** 6%   **(d)** 20%

3 Michelle earns £16 000 per year.
   She receives a salary increase of 3%.
   Find her new salary.

4 A company cuts its wage bill by 15%.
   Its wage bill before the cut was £4 600 000.
   What was the wage bill after the cut?

5 Amy earns £360 per week.
   She pays 15% income tax.
   Calculate Amy's weekly take-home pay.

6 A furniture shop has a sale.

40% off

Copy and complete the table to find the sale price of these articles.

Item	Original Price (£)	Reduction (£)	Sale price (£)
Bookcase	180		
Dining table and chairs	840		
Sofa	440		
Display unit	260		

You may use your calculator for questions **7** to **12**.

7 Increase £84 by these percentages.
   (a) 14%   **(b)** 36%
   (c) 9%   **(d)** 72%

8 Decrease £428 by these percentages.
   (a) 23%   **(b)** 37%
   (c) 7%   **(d)** 69%

9 The value of a car fell by 12% in the first year.
It cost £16 800 when new.
What was its value after 1 year?

10 8% more people passed an examination in 2010 than in 2009.
6400 passed in 2009.
How many passed in 2010?

11 In 1971 there were 902 000 births in the UK.
By 1981 the birth rate had fallen by 19%.
Calculate the number of births in 1981.
Give your answer to the nearest thousand.

12 An antique increased in value by 180% in five years.
It was worth £240 at the start of the five years.
What was it worth at the end of the five years?

Exercise 12.4H

Do not use your calculator for questions **1** to **5**.

1 Calculate £4 as a percentage of £50.

2 Calculate 8 metres as a percentage of 40 metres.

3 Calculate 90p as a percentage of £3.

4 Graeme earns £50 per week.
He receives a pay increase of £2 per week.
Calculate his pay increase as a percentage of £50.

5 A computer originally costing £600 is reduced by £90.
Calculate the reduction as a percentage of the original price.

You may use your calculator for questions **6** to **10**.

6 Calculate £64 as a percentage of £400.

7 Calculate 180 kg as a percentage of 400 kg.

8 The number of houses built in a town in 2008 was 480.
This increased by 168 in 2009.
Find the increase as a percentage of 480.

9 The value of an investment fell from £4800 in 2004 to £4200 in 2009.
Find the reduction as a percentage of £4800.

10 In 2010 the national minimum wage was raised from £5.80 to £5.93.
Calculate the increase as a percentage of £5.80.
Give your answer to the nearest whole number.

Rotation

Exercise 13.1H

1 Describe the rotation symmetry of these shapes.

(a)

(b)

(c)

(d)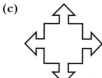

2 Look at this regular pentagon.

How many lines of symmetry does it have?
What is its order of rotation symmetry?

3 This triangle has rotation symmetry of order 3.

Draw a different triangle that has no rotation
symmetry.

4 Make a copy of each of these shapes.
If there is rotation symmetry, show the centre
and state the order of rotation symmetry.

(a)

(b)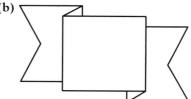

(c)

(d)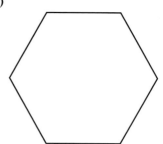

(e)

Exercise 13.2H

1 Copy this grid.
Shade more squares so that the pattern has
rotation symmetry of order 2.

2 Copy this grid.
Shade more squares so that the pattern has
rotation symmetry of order 4.

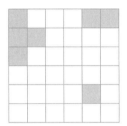

3 Shade squares on a 3 × 3 grid so that your
pattern has rotational symmetry of order 2 but
no reflection symmetry.

4 Copy this diagram.

Complete it so that it has rotation symmetry of
order 4.

5 Copy this diagram.

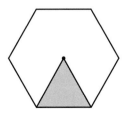

Complete it so that it has rotation symmetry of
order 3.

Exercise 13.3H

1 Copy the diagram.
   (a) Rotate trapezium A through 180° about the
   origin.
   Label it B.
   (b) Rotate trapezium A through 90° clockwise
   about the point (0, 1).
   Label it C.
   (c) Rotate trapezium A through 90°
   anticlockwise about the point (−1, 1).
   Label it D.

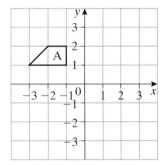

2 Copy the diagram.
   (a) Rotate flag A through 90° clockwise about
   the origin.
   Label it B.
   (b) Rotate flag A through 90° anticlockwise
   about the point (1, −1).
   Label it C.
   (c) Rotate flag A through 180° about the point
   (0, −1).
   Label it D.

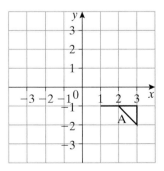

3 Draw *x*- and *y*-axes from −4 to 4.

  (a) Draw a triangle with vertices at (1, 1), (2, 1) and (2, 3).
    Label it A.

  (b) Rotate triangle A through 90° anticlockwise about the origin.
    Label it B.

  (c) Rotate triangle A through 180° about the point (2, 1).
    Label it C.

  (d) Rotate triangle A through 90° clockwise about the point (−2, 1).
    Label it D.

4 Copy the diagram.

  (a) Rotate triangle T through
    90° anticlockwise about the
    point (5, 5).
    Label it A.

  (b) Rotate triangle T through 180° about the
    point (0, 3).
    Label it B.

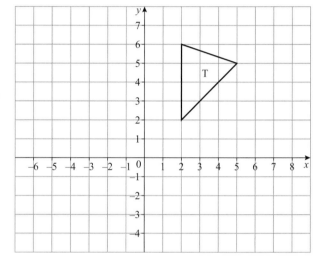

5 Copy the diagram.

  (a) Rotate shape T through 90° anticlockwise
    about the point (−5, −2).
    Label it A.

  (b) Rotate shape T through 90° clockwise
    about the point (0, −1).
    Label it B.

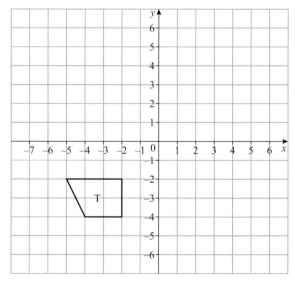

Exercise 13.4H

1 Describe fully the transformation that maps
   (a) trapezium A on to trapezium B.
   (b) trapezium A on to trapezium C.
   (c) trapezium A on to trapezium D.

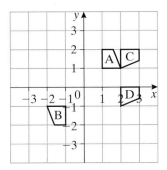

2 Describe fully the transformation that maps
   (a) flag A on to flag B.
   (b) flag A on to flag C.
   (c) flag A on to flag D.

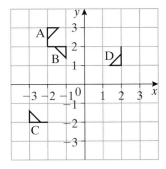

3 Describe fully the transformation that maps
   (a) triangle A on to triangle B.
   (b) triangle A on to triangle C.
   (c) triangle A on to triangle D.
   (d) triangle A on to triangle E.
   (e) triangle A on to triangle F.
   Hint: Some of these transformations are
   reflections.

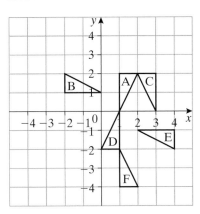

Chapter 14 Estimation

Exercise 14.1H

1. Round each of these numbers to 1 significant figure.
   - (a) 851
   - (b) 276
   - (c) 8432
   - (d) 801
   - (e) 942
   - (f) 38.75
   - (g) 6.23
   - (h) 71.4
   - (i) 99.83
   - (j) 568 790

2. Round each of these numbers to 1 significant figure.
   - (a) 6.8
   - (b) 7.32
   - (c) 48.3
   - (d) 0.63
   - (e) 0.89
   - (f) 4382
   - (g) 587
   - (h) 0.063
   - (i) 0.0077
   - (j) 0.000 409

3. 641 people attended a firework display. How many is this to 1 significant figure?

4. (a) Jenny measured a piece of wood as 2.84 m long. What is this to 1 significant figure?
   (b) Ben measured the wood as 284 cm. What is this to 1 significant figure?

For questions **5** to **10**, show the approximations that you use to get your estimate.

5. Alan bought 22 cans of drink at 39p each. Estimate how much he paid in total.

6. Tickets for a firework display are £18.50 each. 4125 tickets are sold. Estimate the total value of the sales.

7. A rectangle measures 3.9 cm by 8.1 cm. Estimate its area.

8. A rectangle is 4.8 cm wide and its area is 32.1 cm².
   (a) Estimate the length of the rectangle.
   (b) Is the estimate smaller or bigger than the actual length? Explain your answer.

9. Sophie cut up 2.9 m of ribbon into pieces 0.18 m long. Estimate how many pieces she had.

10. Estimate the result of each of these calculations.
    - (a) 6.32×7.12
    - (b) $28.7 \div 6.3$
    - (c) 48.3×32.1
    - (d) $7896 \div 189$
    - (e) 286×0.32
    - (f) 18.9^2
    - (g) $913 \div 196$
    - (h) $4.7 \times 6.2 \times 9.8$

11. Bill wants to turf a rectangular lawn measuring 12.61 metres by 9.78 metres. A company charges £4.75 per square metre to provide and lay the turf. Estimate how much it will cost to turf the lawn.

12. Kate is going to paint the ceiling of the village hall. The hall measures 12.7 m by 7.2 m. A one-litre tin of paint covers 6.5 m². Estimate how many one-litre tins of paint she needs to buy.

Exercise 14.2H

1. Look at these calculations. The answers are all wrong. For each calculation show how you can tell this quickly, without using a calculator to work it out.
   - (a) $15.3 \div -5.1 = 5$
   - (b) $8.7 \times 1.6 = 5.4375$
   - (c) $4.7 \times 300 = 9400$
   - (d) $7.5^2 = 46.25$
   - (e) $5400 \div 9 = 60$
   - (f) $-6.2 \times -0.5 = -93.1$
   - (g) $\sqrt{0.4} = 0.2$
   - (h) $8.5 \times 7.1 = 60.36$

2. Estimate the answer to each of these calculations. Show your working.
   - (a) 93×108
   - (b) 0.61^2
   - (c) $-19.6 + 5.2$

3. Use estimates to calculate a rough cost for each of these.
   - (a) Three DVDs at £17.99
   - (b) 39 cinema tickets at £6.20
   - (c) Five meals at £7.99 and two drinks at £2.10

Enlargement

Exercise 15.1H

1 For each of these shapes:
- copy the shape on to squared paper.
- draw an enlargement of the shape using the scale factor given.

(a) Scale factor 2

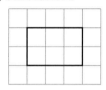

(b) Scale factor 3

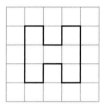

(c) Scale factor 3

(d) Scale factor 2

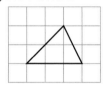

(e) Scale factor 2

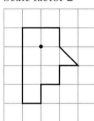

2 Work out the scale factor of enlargement of each of these pairs of shapes.

(a)

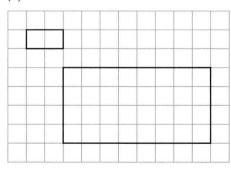

(b)

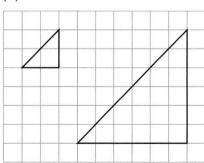

(c)

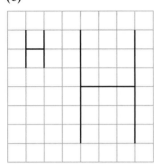

(d)

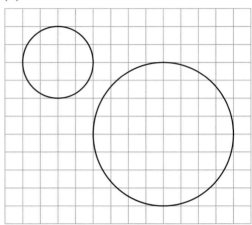

3 For each of these pairs of shapes, is the larger shape an enlargement of the smaller shape?
Give a reason for your answer.

(a)

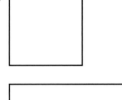

(b)

Exercise 15.2H

1 Copy each of the shapes on to squared paper.
Enlarge each of them by the scale factor given.
Use the dot as the centre of the enlargement.

(a) Scale factor 3

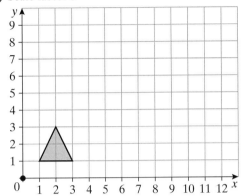

(b) Scale factor 2

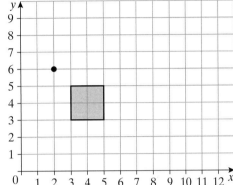

(c) Scale factor 4

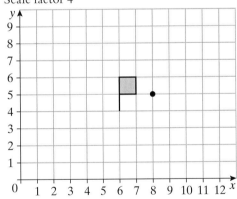

(d) Scale factor 3

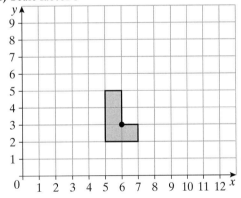

(e) Scale factor 2

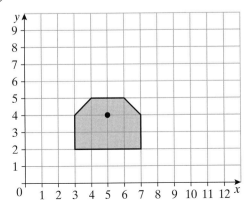

2 Copy each of these diagrams on to squared paper.
For each of these diagrams find
  (i) the scale factor of the enlargement.
  (ii) the coordinates of the centre of the enlargement.

(a)

(b)

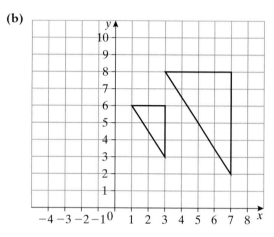

(c)

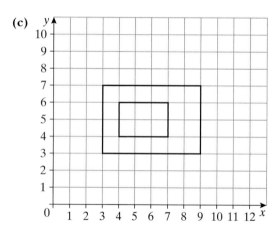

(d)

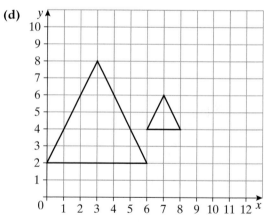

(e)

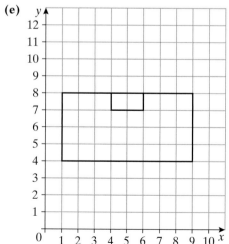

Exercise 15.3H

1 Draw a pair of axes and label them 0 to 6 for both x and y.
  (a) Draw a triangle with vertices at $(0, 6)$, $(3, 6)$ and $(3, 3)$.
  Label it A.
  (b) Enlarge triangle A by scale factor $\frac{1}{3}$, with the origin as the centre of enlargement.
  Label it B.
  (c) Describe fully the transformation that maps triangle B on to triangle A.

2 Draw a pair of axes and label them 0 to 6 for both x and y.
  (a) Draw a triangle with vertices at $(5, 2)$, $(5, 6)$ and $(3, 6)$.
  Label it A.
  (b) Enlarge triangle A by scale factor $\frac{1}{2}$, with centre of enlargement $(3, 2)$.
  Label it B.
  (c) Describe fully the transformation that maps triangle B on to triangle A.

3 Draw a pair of axes and label them 0 to 8 for both x and y.
  (a) Draw a triangle with vertices at $(2, 1)$, $(2, 3)$ and $(3, 2)$.
  Label it A.
  (b) Enlarge triangle A by scale factor $2\frac{1}{2}$, with the origin as the centre of enlargement.
  Label it B.
  (c) Describe fully the transformation that maps triangle B on to triangle A.

4 Draw a pair of axes and label them 0 to 7 for both x and y.
  (a) Draw a trapezium with vertices at $(1, 2)$, $(1, 3)$, $(2, 3)$ and $(3, 2)$.
  Label it A.
  (b) Enlarge triangle A by scale factor 3, with centre of enlargement $(1, 2)$.
  Label it B.
  (c) Describe fully the transformation that maps triangle B on to triangle A.

5 Describe fully the transformation that maps
  (a) triangle A on to triangle B.
  (b) triangle B on to triangle A.
  (c) triangle A on to triangle C.
  (d) triangle C on to triangle A.

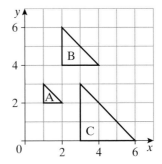

6 Describe fully the transformation that maps
  (a) flag A on to flag B.
  (b) flag B on to flag C.
  (c) flag B on to flag D.
  (d) flag B on to flag E.
  (e) flag F on to flag G.
  Hint: Not all the transformations are enlargements.

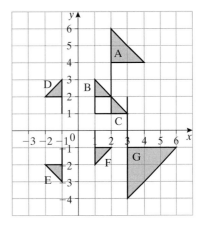

Scatter diagrams and time series

Exercise 16.1H

1 The scatter diagram shows the second–hand
   value of some Ford Focus cars.
   Comment on the results shown by the
   scatter diagram.

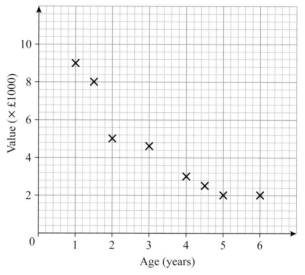

2 The scatter diagram shows some information about
   drivers caught speeding.
   It shows their ages and by how much they were exceeding
   the speed limit.
   Comment on the results shown by the scatter diagram.

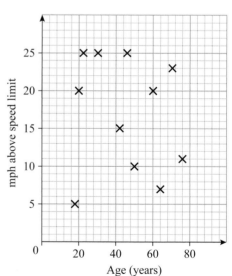

3 This scatter diagram shows the marks gained by ten
students in their mathematics and history exams.
Comment on the results shown by the scatter diagram.

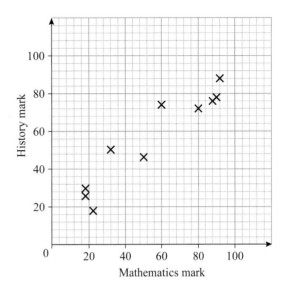

4 Bill grows tomatoes.
As an experiment he divided his land into eight plots.
He used a different amount of fertiliser on each plot.
The table shows the weight of tomatoes he got from each of the plots.

Amount of fertiliser (g/m²)	10	20	30	40	50	60	70	80
Weight of tomatoes (kg)	36	41	58	60	70	76	75	92

(a) Draw a scatter diagram to show this information.
(b) Describe the correlation shown in the scatter diagram.
(c) Draw a line of best fit on your scatter diagram.
(d) What weight of tomatoes should Bill expect to get if he uses 75 g/m² of fertiliser?

5 The table shows the prices and mileages of seven second–hand cars of the same model.

Price (£)	6000	3500	1000	8500	5500	3500	6000
Mileage	27 000	69 000	92 000	17 000	53 000	82 000	43 000

(a) Draw a scatter diagram to show this information.
(b) Describe the correlation shown in the scatter diagram.
(c) Draw a line of best fit on your scatter diagram.
(d) Use your line of best fit to estimate
    (i) the price of this model of car which has covered 18 000 miles.
    (ii) the mileage of this model of car which costs £4000.

6 The heights of ten daughters, all aged 20, and their fathers are given in the table below.

Height of father (cm)	167	168	169	171	172	172	174	175	176	182
Height of daughter (cm)	164	166	166	168	169	170	170	171	173	177

(a) Draw a scatter diagram to show this information.
(b) Describe the correlation shown in the scatter diagram.
(c) Draw a line of best fit on your scatter diagram.
(d) Use your line of best fit to estimate the height of a 20-year-old daughter whose father is 180 cm tall.

Exercise 16.2H

1 The table shows the number of people visiting a cinema complex each day for a 4-week period.

	Su	**M**	**Tu**	**W**	**Th**	**F**	**Sa**
Week 1	2135	1051	1604	2015	1752	1854	3045
Week 2	2614	1252	1527	1927	1813	1927	2984
Week 3	2542	1306	1614	1852	1782	2016	3257
Week 4	2968	1422	1827	2094	1905	2283	3371

(a) Plot a time-series graph of these figures.
(b) Describe the trend over this period.

2 The table shows the number of letters delivered by a postman each day for a 4-week period.

	M	**Tu**	**W**	**Th**	**F**	**Sa**
Week 1	201	205	304	196	316	92
Week 2	192	216	296	142	301	115
Week 3	185	205	310	185	298	106
Week 4	170	186	284	135	277	72

(a) Plot a time-series graph of these figures.
(b) Describe the trend over this period.

3 These figures show the number of mobile text messages sent by the Sarah over a period of 3 years.

	1st quarter	2nd quarter	3rd quarter	4th quarter
Year 1	210	192	361	242
Year 2	262	352	572	391
Year 3	354	529	834	682

Describe the trend over this period.

4 The graph shows a shop's quarterly sales of ice-cream.

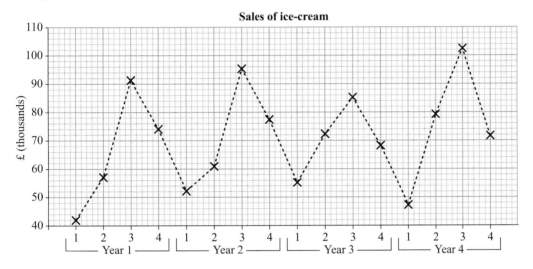

(a) Comment on the seasonal variation shown on the graph.
(b) Which summer had dissapointing sales?

Exercise 17.1H

1 Draw the graph of $y = 3x$ for values of x from -3 to 3.

2 Draw the graph of $y = x + 2$ for values of x from -4 to 2.

3 Draw the graph of $y = 4x + 2$ for values of x from -3 to 3.

4 Draw the graph of $y = 2x - 5$ for values of x from -1 to 5.

5 Draw the graph of $y = -2x - 4$ for values of x from -4 to 2.

6 Draw the graph of $y = 3x - 2$ for values of x from -2 to 4.

7 Draw the graph of $y = 6 - 2x$ for values of x from -2 to 4.

8 (a) Draw the graph of the equation $y = 4x + 3$ for values of x from -3 to 3.
   (b) From the graph find the value of x when $y = 10$.
   Give your answer correct to 1 decimal place.

9 (a) Draw the graph of the equation $C = 3n + 25$ for values of n from 0 to 20.
   (b) From the graph find the value of n when $C = 43$.
   Give your answer correct to the nearest whole number.

Exercise 17.2H

1 Draw the graph of $3y = 2x + 6$, for $x = -3$ to 3.

2 Draw the graph of $2x + 5y = 10$.

3 Draw the graph of $3x + 2y = 15$.

4 Draw the graph of $2y = 5x - 8$, for $x = -2$ to 4.

5 Draw the graph of $3x + 4y = 24$.

6 Draw the graph of $3x + 4y = 12$.

7 Draw the graph of $2x + 6y = 12$.

8 Draw the graph of $2y = 3x - 4$, for $x = -2$ to 4.

Exercise 17.3H

1 For each pair of equations below
   • draw the two graphs on the same grid.
   • write down the coordinates of the point where the two lines cross.
   • write down the simultaneous equations that you have solved.
   (a) $y = 3x$ and $y = 4x - 2$.
   Use values of x from -1 to 4.
   (b) $y = 2x + 3$ and $y = 4x + 1$.
   Use values of x from -2 to 3.
   (c) $y = x + 4$ and $4x + 3y = 12$.
   Use values of x from -3 to 3.
   (d) $y = 2x + 8$ and $y = -2x$.
   Use values of x from -5 to 1.
   (e) $2y = 3x + 6$ and $3x + 2y = 12$.
   Use values of x from 0 to 4.

2 The sum of two numbers is 10 and the difference 5.
   Call the numbers x and y.
   (a) Write down two equations for x and y.
   (b) Draw the graphs of the equations from part **(a)** for values of x from 0 to 10.
   (c) Use your graph to find the two numbers.

Exercise 17.4H

1 Find the gradient of each of these lines and their y-intercept.

(a) $y = 4x + 2$ **(b)** $y = 5 - 3x$

(c) $y + 3 = 2x$ **(d)** $y - 5x = 4$

(e) $y + 5x = 4$ **(f)** $2x + 3y = 6$

(g) $y + \frac{1}{2}x = 3$ **(h)** $3y = 2x - 7$

(i) $2x - 3y = 8$ **(j)** $5x + 4y = 20$

2 On the same diagram, sketch and label the graphs of these three equations.

You do not need to do an accurate plot.

(a) $y = -x + 6$ **(b)** $y = 2x + 5$

(c) $y = 3x - 5$

3 On the same diagram, sketch and label the graphs of these three equations.

You do not need to do an accurate plot.

(a) $y = -2x + 8$ **(b)** $y = \frac{1}{2}x + 3$

(c) $y = 4x + 1$

4 Write down equations of lines parallel to each of these lines.

(a) $y = 2x - 5$ **(b)** $y = -4x + 6$

(c) $y = 8 - 3x$ **(d)** $y = \frac{1}{4}x + 3$

(e) $y = 8 - \frac{1}{2}x$ **(f)** $3x + y = 7$

(g) $x + 5y = 10$ **(h)** $3x - 2y = 12$

Exercise 17.5H

For each of questions **1** to **5**, solve the inequality and show the solution on a number line.

1 $x - 2 > 1$

2 $x + 1 < 3$

3 $3x - 2 \geqslant 7$

4 $2x + 1 \leqslant 6$

5 $3x - 6 \geqslant 0$

For each of questions **6** to **15**, solve the inequality.

6 $7 \leqslant 2x - 1$

7 $5x < x + 12$

8 $4x \geqslant x + 9$

9 $4 + x < 0$

10 $3x + 1 \leqslant 2x + 6$

11 $2(x - 3) > x$

12 $5(x + 1) > 3x + 10$

13 $7x + 5 \leqslant 2x + 30$

14 $5x + 2 < 7x - 4$

15 $3(3x + 2) \geqslant 2(x + 10)$

Exercise 18.1H

1 Which of these triangles are a pair of congruent triangles?

(a)

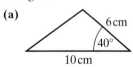
6 cm
40°
10 cm

(b)

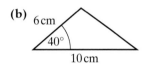
6 cm
40°
10 cm

(c)

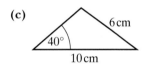
6 cm
40°
10 cm

(d)

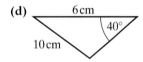
6 cm
40°
10 cm

2 Which of these triangles are congruent to triangle A?

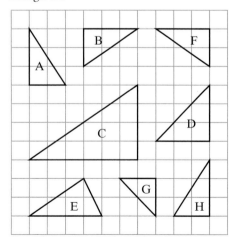

Exercise 18.2H

1 Draw a pair of axes and label them −2 to 6 for x and y.
  (a) Draw a triangle with vertices at (1, 1), (1, 2), and (4, 1).
    Label it A.
  (b) Translate triangle A by vector $\begin{pmatrix} 1 \\ 3 \end{pmatrix}$.
    Label it B.
  (c) Translate triangle A by vector $\begin{pmatrix} -3 \\ 4 \end{pmatrix}$.
    Label it C.
  (d) Translate triangle A by vector $\begin{pmatrix} -2 \\ -3 \end{pmatrix}$.
    Label it D.

2 Draw a pair of axes and label them −3 to 5 for x and y.
  (a) Draw a triangle with vertices at (2, 1), (2, 3) and (3, 1). Label it A.
  (b) Translate triangle A by vector $\begin{pmatrix} 2 \\ 1 \end{pmatrix}$.
    Label it B.
  (c) Translate triangle A by vector $\begin{pmatrix} -5 \\ -3 \end{pmatrix}$.
    Label it C.
  (d) Translate triangle A by vector $\begin{pmatrix} 2 \\ -4 \end{pmatrix}$.
    Label it D.

3 Describe the transformation that maps
  (a) triangle A on to triangle B.
  (b) triangle A on to triangle C.
  (c) triangle A on to triangle D.
  (d) triangle B on to triangle D.

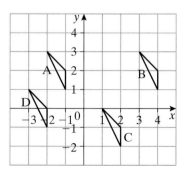

4 Describe the transformation that maps
   (a) shape A on to shape B.
   (b) shape A on to shape C.
   (c) shape A on to shape D.
   (d) shape D on to shape E.
   (e) shape A on to shape F.
   (f) shape E on to shape G.
   (g) shape B on to shape H.
   (h) shape H on to shape F.
   Hint: Not all the transformations are
   translations.

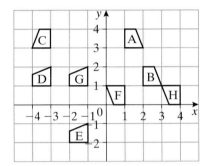

Exercise 18.3H

1 Draw a set of axes. Label the x-axis from 0 to 10
   and the y-axis from −3 to 5.
   Plot the points (1, 2), (1, 4) and (4, 4) and join
   them to form a triangle.
   Label it A.
   (a) Reflect triangle A in the line $x = 5$.
       Label the image B.
   (b) On the same grid, reflect triangle B in the
       line $y = 1$.
       Label the image C.
   (c) Describe fully the single transformation that
       is equivalent to a reflection in the line $x = 5$
       followed by a reflection in the line $y = 1$.

2 Draw a set of axes. Label the x-axis from 0 to 9
   and the y-axis from −4 to 4.
   Plot the points (1, 3), (3, 3) and (3, 2) and join
   them to form a triangle.
   Label it D.
   (a) Rotate triangle D through 90° clockwise
       about the point (0, 0).
       Label the image E.
   (b) On the same grid, rotate triangle E through
       180° about the point (5, −2).
   (c) Describe fully the single transformation that
       is equivalent to a rotation through 90°
       clockwise about the point (0, 0) followed by a
       rotation through 180° about the point (5, −2).

3 Draw a set of axes. Label the x-axis from −5 to 5
   and the y-axis from 0 to 6.
   Plot the points (−3, 1), (0, 1) and (−2, 3) and join
   them to form a triangle.
   Label it A.
   (a) Translate triangle A by the vector $\begin{pmatrix} 4 \\ 0 \end{pmatrix}$.
       Label the image B.
   (b) Translate triangle B by the vector $\begin{pmatrix} -3 \\ 2 \end{pmatrix}$.
       Label the image C.
   (c) Describe fully the single transformation that
       is equivalent to a translation by the vector $\begin{pmatrix} 4 \\ 0 \end{pmatrix}$
       followed by a translation by the vector $\begin{pmatrix} -3 \\ 2 \end{pmatrix}$.

4 Draw a set of axes. Label the x-axis from −6 to 6
   and the y-axis from 0 to 6.
   Plot the points (1, 1), (2, 1) and (1, 3) and join
   them to form a triangle.
   Label it D.
   (a) Enlarge triangle D with scale factor 2 and
       centre of enlargment (0, 0).
       Label the image E.
   (b) Translate triangle E by the vector $\begin{pmatrix} 1 \\ -2 \end{pmatrix}$.
       Label it F.
   (c) Describe fully the single transformation that
       is equivalent to an enlargement with scale
       factor 2 and centre (0, 0) followed by a
       translation by the vector $\begin{pmatrix} 1 \\ -2 \end{pmatrix}$.

In questions **5** to **10**, carry out the transformations on the triangle ABC where A is (1, 1), B is (1, 3) and C is (2, 3).
Describe fully the single transformation equivalent to each of these sets of transformations.

5 Reflection in the line $y = 0$ followed by rotation through 180° about the point (3, 0).

6 Translation by the vector $\begin{pmatrix} 0 \\ -3 \end{pmatrix}$ followed by enlargement with scale factor 2 and centre the origin.

7 Reflection in the line $y = x$ followed by rotation through 90° anticlockwise about the origin.

8 Translation by vector $\begin{pmatrix} 3 \\ 1 \end{pmatrix}$ followed by rotation through 90° clockwise about the point (5, 1).

9 Rotation through 180° about the point (1, 1) followed by enlargement with scale factor 3 and centre (1, 1).

10 Reflection in the y-axis followed by rotation through 180° about the point (−1, 1) followed by reflection in the line $y = 2$.

Unit C Contents

Two-dimensional representation of solids

Exercise 1.1H

1 Match these nets and shapes.
There is no net for one of them.

A   B

C   D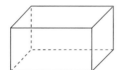

E

(a)   **(b)**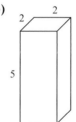

(c)   **(d)**

2 Here are some shapes made out of squares.

A   B   C   D

Which ones will not fold up to make an open box?

Exercise 1.2H

1 Draw a net for each of these cuboids on squared paper. All lengths are in centimetres.

(a)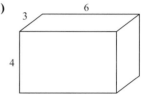

(b)

(c)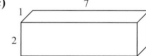

2 Which of these nets are a net for a cuboid?
They are drawn accurately on squared paper.

(a)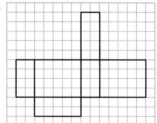

(b)

(c)

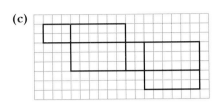

3 For a square-based pyramid, state
  (a) the number of edges.
  (b) the number of vertices.
  (c) the number of faces.

4 When this net is folded to make a cuboid
  (a) which point or points will meet with
        (i) point M?   **(ii)** point A?
  (b) which edge will meet with the edge KJ?

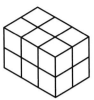

5 A packing case for a lawn mower is a cuboid measuring 80 cm by 60 cm by 40 cm.
It has a top.
Draw a net of the packing case on squared paper.
Use a scale of 1 cm to 20 cm.

6 A seed tray is 30 cm long, 20 cm wide and 5 cm high.
It has no top.
Draw a net for the tray on squared paper.
Use a scale of 1 cm to 5 cm.

7 The box containing Tony's new television is a closed cuboid.
The base is a rectangle of sides 60 cm by 50 cm, and it is 30 cm high.
Draw a net for the box on squared paper.
Use a scale of 1 cm to 20 cm.

Exercise 1.3H

1 Make an isometric drawing of this shape.

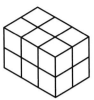

2 Make an isometric drawing of a cuboid with a base 3 cm by 4 cm and 2 cm high.

3 Make an isometric drawing of each of these shapes.

(a)   **(b)**

(c)   **(d)**

(e)   **(f)**

4 Make an isometric drawing of this solid shape.
Use a scale of 1 centimetre to represent 5 metres.

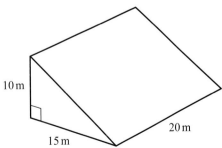

10 m

15 m

20 m

Exercise 1.4H

1 Draw the plan view, front elevation and side elevation of each of these objects.

(a)

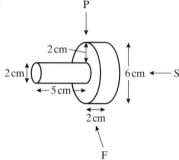

(b)

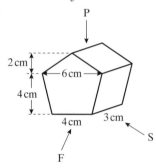

(c)

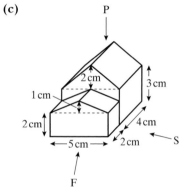

(d)

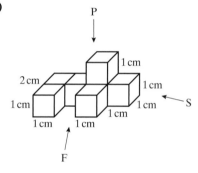

2 A shape is made from six centimetre cubes.
Here are the plan view and the front and side elevations.

Plan   Front   Side

Make an isometric drawing of the shape.

Chapter 2

Probability 1

Exercise 2.1H

1 Choose the best probability word from those below to complete these sentences.

Impossible   Unlikely   Evens
Likely   Certain

(a) It is .... that the next baby born in Devon will be a boy.
(b) It is .... that there will be snow at the South Pole next week.
(c) It is .... that you will get more than 1 when you throw an ordinary dice.

2 Copy this scale.

Impossible Unlikely   Evens   Likely   Certain

Put arrows to show the chance of each of the following events happening.
(a) Water will come out when you turn on the tap.
(b) It will be dark tonight.
(c) Christmas Day falls on a Sunday.

3 There are 20 marbles in a tin. There are 3 red marbles, 10 green marbles and 7 yellow marbles. A marble is taken out without looking. Choose the correct probability word to complete these sentences.
(a) It is .... that the marble is red.
(b) It is .... that the marble is blue.
(c) It is .... that the marble is green.

4 For each part of this question, write numbers for all the cards to make the following statements true when a card is turned over.

(a) There is an even chance that it is 4.
(b) It is impossible to be 4.
(c) It is unlikely to be 4.
(d) It is certain to be 4.

Exercise 2.2H

1 These six cards are laid face down and mixed up. Then a card is picked.

Copy this probability scale.

Impossible   Certain
   0   0.5   1

Put arrows to show the probability of each of the following events happening.
The number on the card is
(a) 3.
(b) less than 6.
(c) an even number.

2 There are ten pens in a bag.
Two are red, seven are blue and one is black.
A pen is taken out without looking.
(a) Use a probability word to complete this sentence.
It is .... that the pen is blue.
(b) Copy this probability scale.

  0   0.2   0.4   0.6   0.8   1.0

Mark the probability of each of these statements on your scale.
Use arrows and label them R, G and B.
R: The pen is red.
G: The pen is green.
B: The pen is blue.

3 This spinner is fair.
Calculate the probability of it landing on
(a) 3.
(b) an even number.
(c) 4.

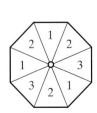

4 An ordinary dice is thrown.
Find the probability of getting
(a) an odd number.
(b) a 4.

5 Mary has ten pens.
Three are black, five are blue and two are green.
She takes a pen without looking.
Giving your answer as a decimal, calculate the
probability that the pen is
(a) blue.
(b) green.
(c) not black.

6 These playing cards are laid face down and
shuffled.
A card is chosen at random.

Giving your answer as a fraction, calculate the
probability that the card is
(a) grey. **(b)** a 7. **(c)** not a 2.

7 Kirsty has eight T-shirts.
She takes one without looking.
How many of her T-shirts are black when
(a) there is an even chance of her getting a black
one?
(b) it is impossible for her to get a black one?

8 There are 12 counters in a bag.
The probability scale shows the probabilities of
choosing the different colours of counter.

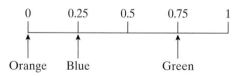

How many counters of each colour are there in
the bag?

9 Gill has eight sweets in a bag.
Three of them are yellow and the other five are
pink.
(a) She takes a sweet without looking.
Which colour is she more likely to get?
(b) Gill takes some sweets out of the bag.
There is now an even chance of getting a
pink sweet.
How many sweets of each colour has she
taken?

10 Jack has ten pairs of socks.
Three of these pairs are black.
He takes a pair of socks out of his drawer without
looking.
Draw a probability scale numbered 0 to 1, as in
question **2**.
Mark the probability of each of these statements
on your scale.
(a) The socks are black.
(b) The socks are not black.

11 The probability is $\frac{3}{8}$ that when three coins are
tossed, you get two heads and a tail.
What is the probability that you don't get two
heads and a tail?

12 There are five yellow counters, two red counters
and eight black counters in a bag.
A counter is taken without looking.
What is the probability that it is
(a) yellow? **(b)** red? **(c)** not red?

13 (a) Copy and complete this table to show the
total score when two spinners numbered 1 to
5 are spun.

	Spinner 1				
	1	**2**	**3**	**4**	**5**
1					
2				6	
3			5		
4					
5					

Spinner 2

(b) What is the smallest possible score?

(c) How many possible ways are there for the two spinners to land?

(d) What is the probability of getting a score of 7?

14 Sita and Ali are getting new bikes for their birthday.

The bikes can be black, silver, red, or green.

(a) Copy and complete this table to show all the possible combinations of colours Sita and Ali's bikes could be.

Two have been done for you.

Sita's bike	Ali's bike
B	B
B	S

(b) Sita says 'The probability that we will both get silver bikes is $\frac{1}{16}$.'

Why might Sita be wrong?

 15 Mr Andrews is choosing his new kitchen.

He can choose beige or white for the cabinets.

He can choose laminate, granite or slate for the worktop.

If he chooses completely randomly, what is the probability that he will choose

(a) white cabinets with a granite worktop?

(b) a kitchen with beige cabinets?

 16 Mike, Ivette and Danny are each buying a new mobile.

The only handsets available to them are made by either Nokia or Samsung.

If they each choose at random, what is the probability they all choose Samsung handsets?

 17 In the game of Monopoly, you throw two dice and your score is the sum of the two numbers.

(a) To land on Old Kent Road, Rafael needs to score 2.

What is the probability that Rafael will land on Old Kent Road on his next go?

(b) If Kellie scores 9, she will land on King's Cross Station.

What is the probability that Kellie lands on King's Cross Station on her next go?

(c) If Ishmael scores 12, he will land on GO.

What is the probability that Ishmael will not land on GO on his next go?

Exercise 2.3H

1 Tim tests a spinner he has made to check it is fair.
   He spins it 500 times.
   Here are his results.

Number on spinner	Frequency
1	58
2	73
3	103
4	124
5	142

(a) For this spinner, calculate the experimental
   probability of obtaining
   (i) a 5.　　**(ii)** a 1.
(b) For a fair spinner, calculate, as a decimal, the
   probability of scoring
   (i) a 5.　　**(ii)** a 1.
(c) Do your answers suggest that the spinner is
   fair? Give your reasons.

2 Nikki spins an eight-sided spinner 200 times and
   records the number of times each score appears.

Number on spinner	Frequency
1	27
2	24
3	23
4	22
5	24
6	26
7	25
8	29

(a) Do you think that Nikki's spinner is fair?
   Give a reason for your answer.
(b) What is the probability that the next spin
   will show a 7?

3 Simon has a biased dice.
   He throws it 500 times and gets these results.

Number	1	2	3	4	5	6
Frequency	20	52	73	158	82	115

Use Simon's results to estimate the probability of
getting a 4 with this dice.

4 Jo throws the same dice as Simon.
   She throws it 50 times and gets 1 five times.
   Explain how her result for throwing a 1 is
   different to Simon's and why this can happen.

5 Sunil made a spinner numbered 0, 1, 2, 3, 4 and
   he tested it 750 times.
   The results are shown in this table.

Number on spinner	Frequency
0	187
1	149
2	236
3	78
4	100

What is the probability that the next spin will
show
(a) 0?　　　　　　(b) an odd number?

6 Irina surveyed the favourite crisp flavour of
   people in her class. Here are her results.

Flavour	Frequency
Plain	7
Chicken	5
Cheese and Onion	12
BBQ	2
Other	4

Calculate an estimate of the probability that the
next person she asks will like cheese and onion
best.

7 Scott makes a note of when his train arrives
each day.
Here are his results for the last 50 days.

Time	Number of days
Early	7
On time	5
Less than 5 minutes late	28
5 minutes late or more	10

Estimate the probability that tomorrow his train
will be
(a) early.   **(b)** late.

8 Jane carries out a survey to see how the students
at her school travel to school.
The results are shown in the table.

Car	Walk	Cycle	Bus	Train	Total
41	47	33	62	17	200

Use this table to estimate the probability that a
student in Jane's school travels by
(a) bus.
(b) cycle.

9 Lizzie and Janna play tennis against each other
each week.
Lizzie won 20 of their last 50 matches and Janna
won 30.
Use these results to estimate the probability that
Lizzie will win their match next week.

Perimeter, area and volume 1

Exercise 3.1H

1 Find the perimeter of each of these shapes.

(a)

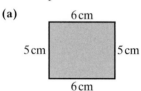

6 cm

5 cm   5 cm

6 cm

(b)

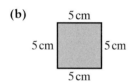

5 cm

5 cm   5 cm

5 cm

(c)

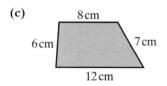

8 cm

6 cm   7 cm

12 cm

(d)

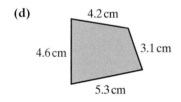

4.2 cm

4.6 cm   3.1 cm

5.3 cm

(e)

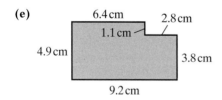

6.4 cm   2.8 cm

1.1 cm

4.9 cm   3.8 cm

9.2 cm

2 A square has sides of length 3.7 m.
What is its perimeter?

3 A rectangle has sides of length 4.3 cm and 5.7 cm.
What is its perimeter?

4 The front of a calculator is a rectangle with sides
of length 13.5 cm and 6.5 cm.
What is its perimeter?

5 The top of a filing cabinet is a rectangle with
sides of length 60 cm and 45 cm.
What is its perimeter?

Exercise 3.2H

1 Find the area of each of these shapes.
Give your answers in cm².

(a)

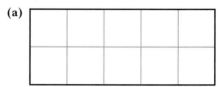

(b)

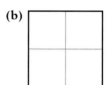

(c)

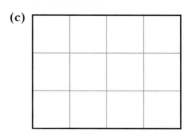

(d)

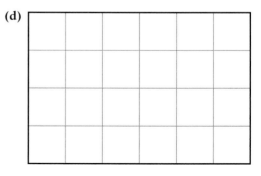

Exercise 3.3H

1 Find the area of each of these rectangles.
   Take care to give the correct units in the answer.

   (a)
   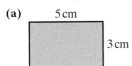
   5 cm
   3 cm

   (b)
   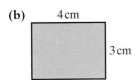
   4 cm
   3 cm

   (c)
   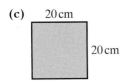
   20 cm
   20 cm

   (d)
   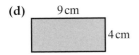
   9 cm
   4 cm

   (e)
   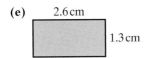
   2.6 cm
   1.3 cm

   (f)
   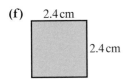
   2.4 cm
   2.4 cm

   (g)
   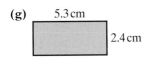
   5.3 cm
   2.4 cm

   (h)
   8.4 cm
   0.6 cm

2 A rectangle measures 5.3 cm by 2.6 cm.
   Find its area.

3 A square has sides of length 3.6 cm.
   Find its area.

4 A rectangle has sides of length 3.25 cm and
   0.95 cm.
   Find its area.

5 A rectangular field measures 35 m by 16 m.
   Find its area.

6 A desk top is a rectangle measuring 1.2 m by
   0.8 m.
   Find its area.

7 A drive is a rectangle measuring 6.5 m by 2.8 m.
   (a) Find the area of the drive.
   To pave the drive it costs £32.50 per square
   metre.
   (b) How much does it cost to pave the drive?

8 A rectangular window measures 1.5 m by 3.5 m.
   (a) Find the area of the window.
   Glass costs £25 per square metre.
   (b) How much does the glass for the window
   cost?

Exercise 3.4H

1 Find the volume of each of these cuboids.

(a)
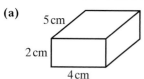
5 cm

2 cm

4 cm

(b)
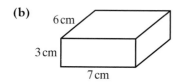
6 cm

3 cm

7 cm

2 A cuboid has a height of 6 cm, a length of 4 cm and a width of 5 cm.
   Find its volume.

3 A cube has edges 4 cm long.
   Find its volume.

4 A cuboid has a square base with sides of length 4.5 cm and is 6 cm high.
   Find its volume.

5 Calculate the volume of a cube with edges 12 m long.

6 A camera box has a base measuring 16 cm by 14 cm and is 12 cm high.
   Calculate its volume.

7 A fish tank has a base measuring 15 cm by 60 cm long and is 25 cm deep.
   Find the volume of the tank.

8 A box of chocolates is a cuboid 12.5 cm long, 6 cm wide and 4.6 cm deep.
   Calculate its volume.

9 A biscuit tin is 12 cm long, 4.5 cm wide and 6.5 cm deep.
   Work out its volume.

10 A piece of wood is 3 m wide, 60 cm long and 2.5 cm thick.
    Find the volume of the wood.
    Hint: Be careful with the units.

11 A box of chocolates is a cuboid 25 cm by 20 cm by 4 cm.
    How many of these can be fitted into a carton 75 cm by 60 cm by 40 cm?

Chapter 4 Measures

Exercise 4.1H

1 Change these units.
  (a) 25 cm to mm
  (b) 24 m to cm
  (c) 1.36 cm to mm
  (d) 15.1 cm to mm
  (e) 0.235 m to mm

2 Change these units.
  (a) 2 m^2 to cm^2
  (b) 3 cm^2 to mm^2
  (c) 1.12 m^2 to cm^2
  (d) 0.05 cm^2 to mm^2
  (e) 2 m^2 to mm^2

3 Change these units.
  (a) 8000 mm^2 to cm^2
  (b) $84\,000 \text{ mm}^2$ to cm^2
  (c) $2\,000\,000 \text{ cm}^2$ to m^2
  (d) $18\,000\,000 \text{ cm}^2$ to m^2
  (e) $64\,000 \text{ cm}^2$ to m^2

4 Change these units.
  (a) 32 cm^3 to mm^3
  (b) 24 m^3 to cm^3
  (c) 5.2 cm^3 to mm^3
  (d) 0.42 m^3 to cm^3
  (e) 0.02 cm^3 to mm^3

5 Change these units.
  (a) $5\,200\,000 \text{ cm}^3$ to m^3
  (b) $270\,000 \text{ mm}^3$ to cm^3
  (c) 210 cm^3 to m^3
  (d) 8.4 m^3 to mm^3
  (e) 170 mm^3 to cm^3

6 Change these units.
  (a) 36 litres to cm^3
  (b) 6300 ml to litres
  (c) 1.4 litres to ml
  (d) 61 ml to litres
  (e) 5400 cm^3 to litres

Exercise 4.2H

1 Copy and complete each of these statements.
  (a) A length given as 4.3 cm, to 1 decimal place, is between .... cm and .... cm.
  (b) A capacity given as 463 ml, to the nearest millilitre, is between .... ml and .... ml.
  (c) A time given as 10.5 seconds, to the nearest tenth of a second, is between .... seconds and .... seconds.
  (d) A mass given as 78 kg, to the nearest kilogram, is between .... kg and .... kg.
  (e) An area given as 5.5 m^2, to 1 decimal place, is between .... m^2 and .... m^2.

2 The number of people attending a football match was given as 24 000 to the nearest thousand. What was the least number of people that could have been at the match?

3 Kerry measures her height as 142 cm to the nearest centimetre.
Write down the two values between which her height must lie.

4 The height of a desk is stated as 75.0 cm to 1 decimal place.
Write down the two values between which its height must lie.

5 Rashid measures the thickness of a piece of plywood as 7.83 mm, to 2 decimal places.
Write down the smallest and greatest thickness it could be.

6 The sides of this triangle are given in centimetres, correct to 1 decimal place.

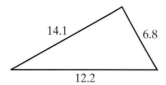

14.1   6.8

12.2

(a) Write down the shortest and longest possible length of each side.
(b) Write down the shortest and longest possible length of the perimeter.

 7 John has two pieces of string.
He measures them as 125 mm and 182 mm, to the nearest millimetre.
He puts the two pieces end to end.
What is the shortest and longest that their combined lengths can be?

 8 Mel and Mary both buy some apples.
Mel buys 3.5 kg and Mary buys 4.2 kg.
Both weights are correct to the nearest tenth of a kilogram.
(a) What is the smallest possible difference between the amounts they have bought?
(b) What is the largest possible difference between the amounts they have bought?

Exercise 4.3H

1 Rewrite each of these statements using sensible values for the measurements.
(a) My mass is 78.32 kg
(b) It takes Katriona 16 minutes and 15.6 seconds to walk to school.
(c) The distance to London from Sheffield is 161.64 miles.
(d) The length of our classroom is 5 metres 14 cm 3 mm.
(e) My water jug hold 3.02 litres.

2 Give your answer to each of these questions to a sensible degree of accuracy.
(a) Estimate the length of this line.

_____

(b) Estimate the size of this angle.

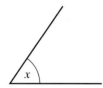

x

(c) A rectangle is 2.3 cm long and 4.5 cm wide.
Find the area of the rectangle.
(d) The volume of a cube is 7 cm³.
Find the length of an edge.
(e) An angle in a pie chart is found by working out $\frac{4}{7} \times 360°$.
Find the angle.
(f) Six friends share £14 between them.
How much does each one get?

Exercise 4.4H

1 Kieran drove 180 kilometres at an average speed of 40 km/h.
   How long did he take?

2 Imogen walked at an average speed of 5 km/h for 3 hours 30 minutes.
   How far did she walk?

3 Patrick drove from his home to London, a distance of 189 miles.
   He took four and a half hours.
   What was his average speed in miles per hour?

4 A car travels at an average speed of 84 km/h.
   How far does it travel in 2.6 hours?

5 It took Peter 4 minutes 10 seconds to run 1000 metres.
   What was his speed in metres per second.

6 Mandy went on a ride on her motor bike.
   Her average speed was 96 km/h and she took 1 hour 40 minutes.
   How far did she travel?

7 A plane takes 3 hours 30 minutes to fly 2611 km.
   What was its average speed?

8 The distance from Lake Louise to Radium Springs is 240 miles.
   Taj travelled at an average speed of 54 mph.
   How long did it take him? Give your answer correct to the nearest minute.

9 A long cycle race covers 275 km.
   The winner took 14 hours 15 minutes.
   What was his average speed? Give your answer correct to 1 decimal place.

10 (a) How long does it take to travel 100 miles at an average speed of 70 mph?
   (b) How much time do you save by travelling 10 mph faster?
       Give your answers correct to the nearest minute.

Exercise 4.5H

1 The density of a rock is 9.3 g/cm^3.
   Its volume is 60 cm^3.
   What is its mass?

2 Calculate the density of a piece of metal with mass 300 g and volume 84 cm^3.
   Give your answer to a sensible degree of accuracy.

3 Calculate the mass of a stone of volume 46 cm^3 and density 7.6 g/cm^3.

4 Copper has a density of 8.9 g/cm^3.
   Calculate the volume of a block of copper of mass 38 g.
   Give your answer to a sensible degree of accuracy.

5 A gas has a mass of 32 kg and occupies a volume of 25 m^3.
   What is its density?
   Give your answer to a sensible degree of accuracy.

6 A small town in America has a population of 235 and covers an area of 35 km^2.
   Find the population density (number of people per square kilometre) of the town.

The area of triangles and parallelograms

Exercise 5.1H

1 Find the area of each of these triangles.

(a)
5 cm
6 cm

(b)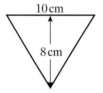
10 cm
8 cm

(c)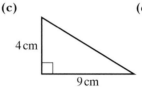
4 cm
9 cm

(d)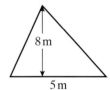
8 m
5 m

(e)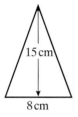
15 cm
8 cm

(f)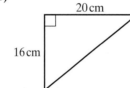
20 cm
16 cm

(g)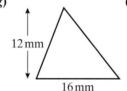
12 mm
16 mm

(h)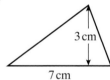
3 cm
7 cm

(i)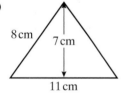
8 cm
7 cm
11 cm

(j)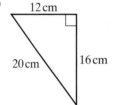
12 cm
20 cm
16 cm

2 A triangle has an area of 15 cm² and a base of 5 cm.

h

Find the perpendicular height *h*.

3   In triangle ABC, AB = 5 cm, BC = 9 cm and AC = 8 cm.
    Angle ABC = 90°.

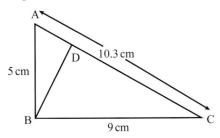

(a) Find the area of the triangle.
(b) Find the perpendicular height BD.

4   The vertices of a triangle are at A(3, 2), B(7, 2) and C(6, 9).
    Find the area of triangle ABC.

Exercise 5.2H

1   Find the area of each of these parallelograms.

(a)
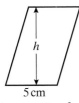
4 cm
7 cm

(b)
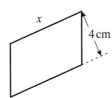
6 cm
9 cm

(c)
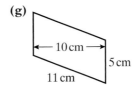
10 cm
2.5 cm

(d)
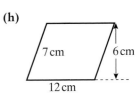
7 m
8 m

(e)
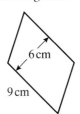
5 cm 6 cm
4 cm

(f)
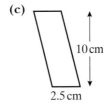
4 cm
5.5 cm

(g)
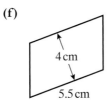
10 cm
5 cm
11 cm

(h)
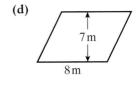
7 cm   6 cm
12 cm

2   Find the length marked with a letter in each of these diagrams.

(a)
h
5 cm
Area = 40 cm²

(b)
x
4 cm
Area = 36 cm²

3 A triangle has a base of 11 cm and a perpendicular height of 8 cm.
A parallelogram has the same area and a base of 10 cm.
What is the perpendicular height of the parallelogram?

4 A parallelogram has an area of 60 cm² and base of 10 cm.
A triangle has the same perpendicular height as the parallelogram and a base of 12 cm.
What is the area of the triangle?

5 A square has an area of 49 cm².
A triangle stands on one side of the square and a parallelogram stands on another.
The triangle and the parallelogram also have areas of 49 cm² each.

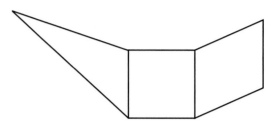

Find the perpendicular height of
(a) the triangle.
(a) the parallelogram.

Chapter 6

Probability 2

Exercise 6.1H

1 The probability that Stacey will go to bed late tonight is 0.2.
   What is the probability that Stacey will not go to bed late tonight?

2 The probability that I will throw a six with a dice is $\frac{1}{6}$.
   What is the probability that I will not throw a six?

3 The probability that it will snow on Christmas Day is 0.15.
   What is the probability that it will not snow on Christmas Day?

4 The probability that someone chosen at random is left-handed is $\frac{3}{10}$.
   What is the probability that they will be right-handed?

5 The probability that United will lose their next game is 0.08.
   What is the probability that United will not lose their next game?

6 The probability that Ian will eat crisps for his lunch is $\frac{17}{31}$.
   What is the probability that he will not eat crisps?

Exercise 6.2H

1 A shop has brown, white and wholemeal bread for sale.
   The probability that someone will choose brown bread is 0.4 and the probability that they will choose white bread is 0.5.
   What is the probability of someone choosing wholemeal bread?

2 A football coach is choosing a striker for the next game.
   He has three players to choose from: Wayne, Michael and Alan.
   The probability that he will choose Wayne is $\frac{5}{19}$ and the probability that he will choose Michael is $\frac{7}{19}$.
   What is the probability that he will choose Alan?

3 A bag contains red, white and blue counters.
   Jill chooses a counter at random.
   The probability that she chooses a red counter is 0.4 and the probability that she chooses a blue counter is 0.15.
   What is the probability that she chooses a white counter?

4 Elaine goes to town by car, bus, taxi or bike.
   The probability that she uses her car is $\frac{12}{31}$, the probability that she catches the bus is $\frac{2}{31}$ and the probability that she takes a taxi is $\frac{13}{31}$.
   What is the probability that she rides her bike into town?

5 A biased five-sided spinner is numbered 1 to 5.
   The table shows the probability of obtaining some of the scores when it is spun.

Score	1	2	3	4	5
Probability	0.37	0.1	0.14		0.22

   What is the probability of getting a 4?

6 A cash bag contains only £20, £10 and £5 notes.
   One note is chosen from the bag at random.
   There is a probability of $\frac{3}{4}$ that it is a £5 note and a probability of $\frac{3}{20}$ that it is a £10 note.
   What is the probability that it is a £20 note?

Exercise 6.3H

1 The probability that United will lose their next game is 0.2.
   How many games would you expect them to lose in a season of 40 games?

2 The probability that it will rain on any day in June is $\frac{2}{15}$.
   On how many of June's 30 days would you expect it to rain?

3 The probability that an eighteen-year-old driver will have an accident is 0.15.
   There are 80 eighteen-year-old drivers in a school.
   How many of them might be expected to have an accident?

4 When Phil is playing chess, the probability that he wins is $\frac{17}{20}$.
   In a competition, Phil plays ten games.
   How many of them might you expect him to win?

5 An ordinary six-sided dice is thrown 90 times.
   How many times might you expect to get
   (a) a 4?
   (b) an odd number?

6 A box contains twelve yellow balls, three blue balls and five green balls.
   A ball is chosen at random and its colour noted.
   The ball is then replaced.
   This is done 400 times.
   How many of each colour might you expect to get?

Exercise 6.4H

1 Pete rolls a dice 200 times and records the number of times each score appears.

Score	1	2	3	4	5	6
Frequency	29	34	35	32	34	36

   (a) Work out the relative frequency of each of the scores.
   Give your answers to 2 decimal places.
   (b) Do you think that Pete's dice is fair?
   Give a reason for your answer.

2 Rory kept a record of his favourite football team's results.

   Win: 32   Draw: 11   Lose: 7

   (a) Calculate the relative frequency of each of the three outcomes.
   (b) Are your answers to part **(a)** good estimates of the probability of the outcome of their next match?
   Give a reason for your answer.

3 In a survey, 600 people were asked which flavour of crisps they preferred.
   The results are shown in the table.

Flavour	Frequency
Plain	166
Salt and vinegar	130
Cheese and onion	228
Other	76

   (a) Work out the relative frequency for each flavour.
   Give your answers to 2 decimal places.
   (b) Explain why it is reasonable to use these figures to estimate the probability of the flavour of crisps that the next person to be asked will prefer.

4 The owner of a petrol station notices that in one day 287 out of 340 people filling their car with petrol spent over £20.
Use these figures to estimate the probability that the next customer will spend
(a) over £20.
(b) £20 or less.

5 Jasmine made a spinner numbered 1, 2, 3, 4 and 5. She tested the spinner to see if it was fair.
The results are shown below.

Score	1	2	3	4	5
Frequency	46	108	203	197	96

(a) Work out the relative frequency of each of the scores.
Give your answers to 2 decimal places.
(b) Do you think that the spinner is fair?
Give a reason for your answer.

6 A box contains yellow, green, white and blue counters.
A counter is chosen from the box and its colour noted.
The counter is then replaced in the box.
The table below gives information about the colour of counter picked.

Colour	Relative frequency
Yellow	0.4
Green	0.3
White	0.225
Blue	0.075

(a) There are 80 counters altogether in the bag.
How many do you think there are of each colour?
(b) What other information is needed before you can be sure that your answers to part (a) are accurate?

Exercise 7.1H

1 Find the circumferences of the circles with these diameters.
   (a) 8 cm   (b) 17 cm
   (c) 39.2 cm   (d) 116 mm
   (e) 5.1 m   (f) 6.32 m
   (g) 14 cm   (h) 23 cm
   (i) 78 mm   (j) 39 mm
   (k) 4.4 m   (l) 2.75 m

2 Find the circumferences of the circles with these radii.
   (a) 3 cm   (b) 25 m
   (c) 56 mm   (d) 4.8 cm
   (e) 5.12 m   (f) 62.4 cm

 3 The diagram shows a bicycle wheel surrounded by a tyre.

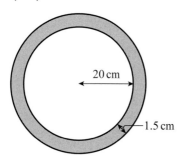

What distance does the wheel travel when it turns one complete circle?

Exercise 7.2H

1 Find the areas of the circles with these radii.
   (a) 17 cm   (b) 23 cm
   (c) 67 cm   (d) 43 mm
   (e) 74 mm   (f) 32 cm
   (g) 58 cm   (h) 4.3 cm
   (i) 8.7 cm   (j) 47 m
   (k) 1.9 m   (l) 2.58 m

2 Find the areas of the circles with these diameters.
   (a) 18 cm   (b) 28 cm
   (b) 68 cm   (d) 38 mm
   (e) 78 mm   (f) 58 cm
   (g) 46 cm   (h) 6.4 cm
   (i) 7.6 cm   (j) 32 m
   (k) 3.4 m   (l) 4.32 m

3 Find
   (a) the circumference
   (b) the area
   of this circle.
   Leave your answer in terms of π.

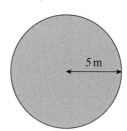

Exercise 7.3H

Find the area of each of these shapes.
Break them down into rectangles and right-angled triangles first.

1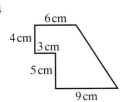

2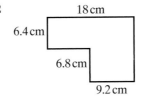

3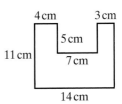

4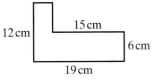

5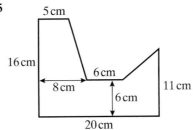

6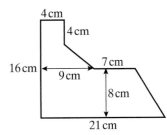

Exercise 7.4H

Find the volume of each of these shapes.

1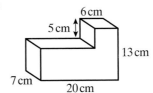

2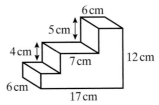

3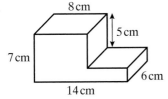

4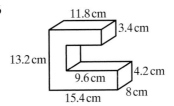

5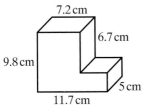

6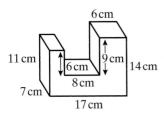

Exercise 7.5H

Find the volume of each of these shapes.

1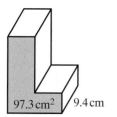
97.3 cm² / 9.4 cm

2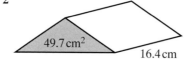
49.7 cm²   16.4 cm

3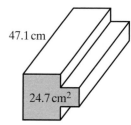
47.1 cm
24.7 cm²

4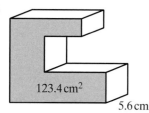
123.4 cm²
5.6 cm

5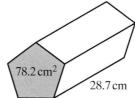
78.2 cm²   28.7 cm

6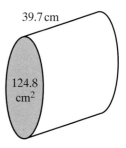
39.7 cm
124.8 cm²

Exercise 7.6H

1 The edges of a cube are 5 cm long.
  Calculate the surface area of the cube.

2 A box is 10 cm high, 5 cm long and 3 cm wide.
  Calculate the surface area of the box.

3 A classroom is 6 metres long, 4 metres wide and 3 metres high.
  Calculate the surface area of the walls.

4 A biscuit tin is a cuboid 12 cm long, 5 cm wide and 6 cm deep and has a lid.
  Calculate the surface area of the tin.

5 This is a sketch of the net for a regular tetrahedron (triangular-based pyramid).

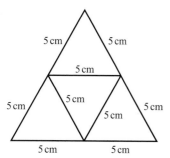

5 cm   5 cm
5 cm
5 cm   5 cm
5 cm   5 cm
5 cm   5 cm
5 cm   5 cm

Construct the net accurately.
Take measurements from your drawing and hence calculate the surface area of the tetrahedron.

6 The volume of cuboid B is twice that of cuboid A. All lengths are in centimetres.

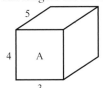

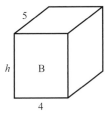

(a) Find h.

(b) Is the surface area of cuboid B twice that of cuboid A?

Show your calculations.

Exercise 7.7H

1 Find the volumes of the cylinders with these dimensions.

(a) Radius 7 cm and height 29 cm

(b) Radius 13 cm and height 27 cm

(c) Radius 25 cm and height 80 cm

(d) Radius 14 mm and height 35 mm

(e) Radius 28 mm and height 8 mm

(f) Radius 0.6 mm and height 5.1 mm

(g) Radius 1.7 m and height 5 m

(h) Radius 2.6 m and height 3.4 m

2 A cylindrical vase has a radius of 9 cm and a height of 20 cm.

Find the volume of the vase in terms of π.

Exercise 7.8H

1 Find the curved surface areas of the cylinders with these dimensions.

(a) Radius 9 cm and height 16 cm

(b) Radius 13 cm and height 21 cm

(c) Radius 27 cm and height 12 cm

(d) Radius 17 mm and height 35 mm

(e) Radius 12 mm and height 6 mm

(f) Radius 3.7 mm and height 63 mm

(g) Radius 1.9 m and height 19 m

(h) Radius 2.7 m and height 4.3 m

2 Find the total surface areas of the cylinders with these dimensions.

(a) Radius 8 cm and height 11 cm

(b) Radius 17 cm and height 28 cm

(c) Radius 29 cm and height 15 cm

(d) Radius 32 mm and height 8 mm

(e) Radius 35 mm and height 12 mm

(f) Radius 3.9 mm and height 45 mm

(g) Radius 0.8 m and height 7 m

(h) Radius 2.9 m and height 1.7 m

3 Find the total surface area of the cylinder with a radius of 8 cm and height of 7 cm.

Give your answer in terms of π.

 4 The roller on a lawn mower is a cylinder of diameter 30 cm and width 40 cm.

It is used to cut a grass verge measuring 80 cm by 50 m.

How many revolutions will the roller make to cut the grass without any overlaps?

Hint: Take care with the units.

Exercise 8.1H

Work these out on your calculator without writing down the answers to the middle stages.
If the answers are not exact, give them correct to 2 decimal places.

1 $\dfrac{7.3 + 8.5}{5.7}$

2 $\dfrac{158 + 1027}{125}$

3 $\dfrac{6.7 + 19.5}{12.2 - 5.7}$

4 $\sqrt{128 - 34.6}$

5 $5.7 + \dfrac{1.89}{0.9}$

6 $(12.6 - 9.8)^2$

7 $\dfrac{8.9}{2.3 \times 5.6}$

8 $\dfrac{15.4}{2.3^2}$

9 $10.9 \times (7.2 - 5.8)$

10 $\dfrac{4.8 + 6.2}{5.2 \times 6.5}$

11 $\dfrac{7.1}{\sqrt{15.3 \times 0.6}}$

12 $\dfrac{3 - \sqrt{2.73 + 5.1}}{4}$

Exercise 8.2H

1 Work out these.
    (a) $\frac{3}{4} + \frac{1}{6}$ (b) $\frac{5}{8} - \frac{2}{7}$ (c) $\frac{5}{9} \times \frac{3}{8}$
    (d) $\frac{7}{16} \div \frac{5}{12}$ (e) $1\frac{4}{5} + 2\frac{3}{4}$ (f) $6\frac{3}{7} - 2\frac{1}{3}$
    (g) $5\frac{3}{5} \times 4$ (h) $4\frac{5}{9} \div 1\frac{1}{6}$

2 Write these fractions in their lowest terms.
    (a) $\frac{40}{125}$ (b) $\frac{28}{49}$ (c) $\frac{72}{192}$
    (d) $\frac{225}{350}$ (e) $\frac{17}{153}$

3 Write these improper fractions as mixed numbers.
    (a) $\frac{120}{72}$ (b) $\frac{150}{13}$ (c) $\frac{86}{19}$
    (d) $\frac{192}{54}$ (e) $\frac{302}{17}$

4 Calculate
    (a) the perimeter of this rectangle.
    (b) the area of this rectangle.

$5\frac{3}{5}$ cm

$2\frac{1}{4}$ cm

Exercise 8.3H

1 Round each of these numbers to 1 significant figure.
(**a**) 8.4   (**b**) 18.36   (**c**) 725
(**d**) 8032   (**e**) 98.3   (**f**) 0.71
(**g**) 0.0052   (**h**) 0.019   (**i**) 407.511
(**j**) 23 095

2 Round each of these numbers to 2 significant figures.
(**a**) 28.7   (**b**) 149.3   (**c**) 7832
(**d**) 46 820   (**e**) 21.36   (**f**) 0.194
(**g**) 0.0489   (**h**) 0.003 61   (**i**) 0.0508
(**j**) 0.904

3 Round each of these numbers to 3 significant figures.
(**a**) 7.385   (**b**) 24.81   (**c**) 28 462
(**d**) 308.61   (**e**) 16 418   (**f**) 3.917
(**g**) 60.72   (**h**) 0.9135   (**i**) 0.004 162
(**j**) 2.236 06

4 At a county cricket match, the attendance was 11 035 men, 6140 women and 3775 juniors.
The local newspaper reported that there were 21 000 people at the match.
Explain this.

5 Over the three days of an agricultural show there were 8600 visitors.
On average, how many visitors were there each day correct to 2 significant figures?

6 The area of England is 50 318 square miles.
1 square mile is approximately 2.59 square kilometres.
What is the area of England, correct to the nearest square kilometre.

7 The 22 members of a club went on a trip together.
The total cost was £716.58.
They each paid the same amount.
How much did each pay?

8 Mike wants to paint the floor of his garage.
The garage is 4.8 m by 6.1 m.
One litre of floor paint covers 2.5 m^2.
How many one-litre tins of paint does he need to buy?

9 A rectangular driveway is measured as 8.21 metres by 3.64 metres.
A company charges £18.75 per square metre to repair driveways.
How much would they charge to repair the driveway?
Give your answer correct to the nearest pound.

10 Work out these.
Give your answers to 3 significant figures.

(**a**) 71×58 (**b**) $\sqrt{46}$

(**c**) $\dfrac{5987}{5.1}$ (**d**) 19.1^2

(**e**) 62.7×8316 (**f**) $\dfrac{5.72}{19.3}$

(**g**) $\dfrac{32}{49.4}$ (**h**) 8152×37

(**i**) $\dfrac{9.35 \times 4.1}{48.5}$ (**j**) $\dfrac{673 \times 0.76}{3.6 \times 2.38}$

Trial and improvement

Exercise 9.1H

1 **(a)** Calculate the value of $x^3 - x$ when
    (i) $x = 4$.   **(ii)** $x = 5$.
    (iii) $x = 4.6$.   **(iv)** $x = 4.7$.
    (v) $x = 4.65$.
  (b) Using your answers to part **(a)**, give the solution of $x^3 - x = 94$, to 1 decimal place.

2 Find a solution, between $x = 2$ and $x = 3$, to the equation $x^3 = 11$.
Give your answer correct to 1 decimal place.

3 **(a)** Show that a solution to the equation
    $x^3 + 3x = 30$ lies between $x = 2$ and $x = 3$.
  (b) Find the solution correct to 1 decimal place.

4 **(a)** Show that a solution to the equation
    $x^3 - 2x = 70$ lies between $x = 4$ and $x = 5$.
  (b) Find the solution correct to 1 decimal place.

5 Find a solution to the equation $x^3 + 4x = 100$.
Give your answer correct to 1 decimal place.

6 Find a solution to the equation $x^3 + x = 60$.
Give your answer correct to 2 decimal places.

7 Find a solution to the equation $x^3 - x^2 = 40$.
Give your answer correct to 2 decimal places.

8 A number, x, added to the square of that number is equal to 1000.
  (a) Write this as an equation.
  (b) Find the number correct to 1 decimal place.

9 The cube of a number minus the number is equal to 600.
Find the number correct to 2 decimal places.

Chapter 10 | Enlargement

Exercise 10.1H

1 The triangles PQR and STU are similar.
Calculate the lengths of ST and SU.

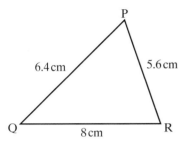

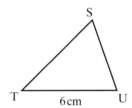

2 The triangles ABC and DEF are similar.
Calculate the lengths of AB and EF.

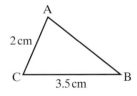

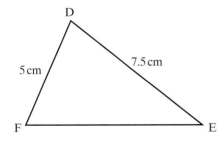

3 Triangles ABC and AXY are similar.
Calculate the lengths of BC and CY.

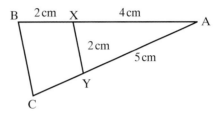

4 Triangles ADE and ABC are similar.
Calculate the lengths of AD and CE.

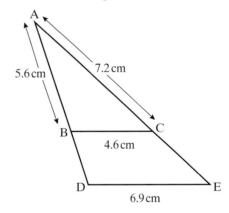

Exercise 10.2H

1 Triangles ABC and PQR are similar.

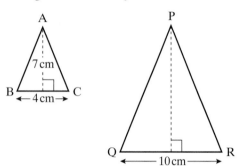

(a) What is the linear scale factor of the enlargement?
(b) Find the height of triangle PQR.
(c) Calculate the area of triangle ABC.
(d) Calculate the area of triangle PQR.
(e) Write down the ratio of the areas.

 2 A small bottle holds 150 ml of liquid.
A similar bottle is twice as tall.
How much liquid does it hold?

 3 Shirley has a poster made from a photo she has taken.
The poster is an enlargement of the photo with a linear scale factor of 8.
The dimensions of the photo are 5 cm by 7 cm.
What is the area of the poster?

4 A vase is 12 cm tall.
Another similar vase is 18 cm tall.
The larger vase has a capacity of 54 cm³.
What is the capacity of the smaller vase?

Exercise 11.1H

1 Simon had a bath.
The graph shows the volume (V gallons) of the water in the bath after t minutes.

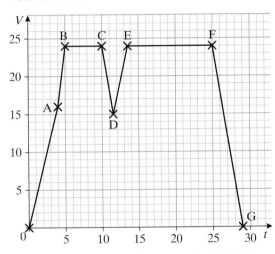

(a) How many gallons of water are in the bath at A?

(b) Simon got in the bath at B and out at F. For how long was he in the bath?

(c) Between O and A, the hot tap is on. How many gallons of water per minute came from the hot tap?

(d) Between A and B, both taps are on. What is the rate of flow of both taps together? Give your answer in gallons/minute.

(e) Describe what happened between C and E.

(f) At what rate did the bath empty? Give your answer in gallons/minute.

2 A printer's charge for printing programmes is worked out as follows.

A fixed charge of £a

+

x pence per programme for the first 1000 programmes

+

80 pence per programme for each programme over 1000

The graph below shows the total charge for printing up to 1000 programmes.

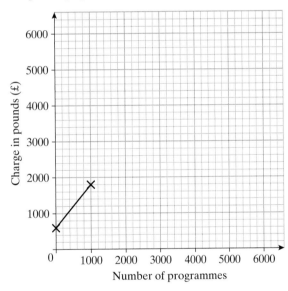

(a) What is the fixed charge, £a?

(b) Calculate x, the charge per programme for the first 1000 programmes.

(c) Copy the graph and add a line segment to show the charges for 1000 to 6000 programmes.

(d) What is the total charge for 3500 programmes?

(e) What is the average cost per programme for 3500 programmes?

3 The graph shows a train journey.

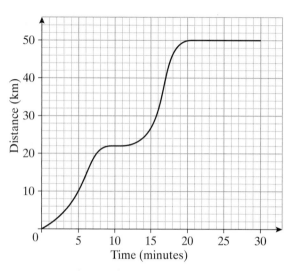

(a) How long did the train journey take?
(b) How far was the train journey?
(c) How far from the start was the first station?
(d) How long did the train stop at the first station?
(e) When was the train travelling fastest?

 4 Water is poured into each of these containers at a constant rate until they are full.

(a)  (b)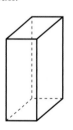

(c) 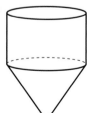 (d)

These graphs show the depth of water (*d*) against time (*t*).
Choose the most suitable graph for each container.

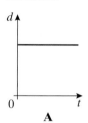
A

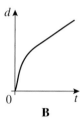
B

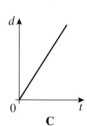
C

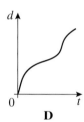
D

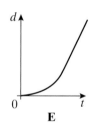
E

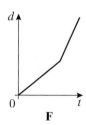
F

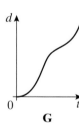
G

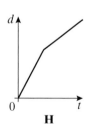
H

5 A mobile phone company offers its customers the choice of two price plans.

	Plan A	Plan B
Monthly subscription	£10.00	£s
Free talk time per month	60 minutes	100 minutes
Cost per minute over the free talk time	*a* pence	35 pence

The graph shows the charges for Plan A and the charges for Plan B up to 100 minutes.

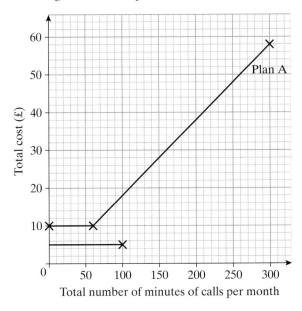

(a) Find the monthly subscription for Plan B (£s).
(b) Shamir uses price Plan B.
He uses the phone for 250 minutes per month.
How much does it cost him?
(c) Copy the graph and add a line to show the charges for Plan B for 100 to 250 minutes.
(d) For how many minutes is the cost the same in both price plans?
(e) Which price plan is the cheaper when the time for calls is 220 minutes?
By how much?

6 The table shows the cost of sending parcels.

Maximum weight	Cost
10 kg	£13.85
11 kg	£14.60
12 kg	£15.35
13 kg	£16.10
14 kg	£16.85
15 kg	£17.60

The graph shows the information in the first row in the table.

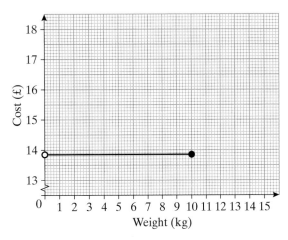

(a) What is the cost of sending a parcel weighing
   (i)   9.6 kg?
   (ii) 10 kg?
   (iii) 10.1 kg?
(b) (i) What is the meaning of the dot at the right of the line?
   (ii) What is the meaning of the circle at the left of the line?
(c) Copy the graph and add lines to show the cost for parcels weighing up to 15 kg.
(d) Hazel posted one parcel weighing 8.4 kg and another weighing 12.8 kg.
What was the total cost?

7 15 000 people attended a football match.
The gates opened 60. minutes before kick off.
The match lasted 90 minutes with a break of
15 minutes at half-time.
200 people left at half-time.
Some people started leaving 10 minutes before
the final whistle, but most stayed to the end.
The grounds were cleared 30 minutes after the
final whistle.
Copy the axes and sketch a graph of the
number of people in the grounds against the time
after the gates opened.

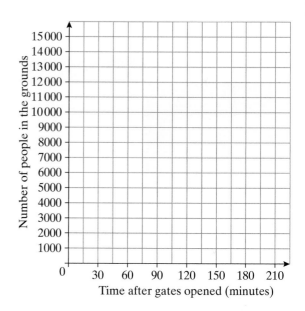

Exercise 11.2H

1 Which of these functions are quadratic?
For each of the functions that is quadratic, state whether the graph is ∪-shaped or ∩-shaped.

(a) $y = x^2 + 7$ (b) $y = 2x^3 + x^2 - 4$ (c) $y = x^2 + x - 7$ (d) $y = x(6 - x)$

(e) $y = \dfrac{5}{x^2}$ (f) $y = x(x^2 + 1)$ (g) $y = 2x(x + 2)$ (h) $y = 5 + 3x - x^2$

2 (a) Copy and complete the table of values for $y = 2x^2$.

x	−3	−2	−1	0	1	2	3
x^2	9					4	
$y = 2x^2$	18					8	

(b) Plot the graph of $y = 2x^2$.
   Use a scale of 2 cm to 1 unit on the x-axis and 1 cm to 1 unit on the y-axis.
(c) Use your graph to
   (i)   find the value of y when $x = -1.8$.
   (ii) solve $2x^2 = 12$.

3 **(a)** Copy and complete the table of values for $y = x^2 + x$.

x	−4	−3	−2	−1	0	1	2	3
x^2			4					9
$y = x^2 + x$								12

(b) Plot the graph of $y = x^2 + x$.
   Use a scale of 2 cm to 1 unit on the x-axis and 1 cm to 1 unit on the y-axis.
(c) Use your graph to
   (i) find the value of y when $x = 1.6$.
   (ii) solve $x^2 + x = 8$.

4 **(a)** Copy and complete the table of values for $y = x^2 − x + 2$.

x	−3	−2	−1	0	1	2	3	4
x^2		4						16
$−x$		2						−4
$+2$		2						2
$y = x^2 − x + 2$		8						14

(b) Plot the graph of $y = x^2 − x + 2$.
   Use a scale of 2 cm to 1 unit on the x-axis and 1 cm to 1 unit on the y-axis.
(c) Use your graph to
   (i) find the value of y when $x = 0.7$.
   (ii) solve $x^2 − x + 2 = 6$.

5 **(a)** Copy and complete the table of values for $y = x^2 + 2x − 5$.

x	−5	−4	−3	−2	−1	0	1	2	3
x^2				4					9
$+2x$				−4					6
$−5$				−5					−5
$y = x^2 + 2x − 5$				−5					10

(b) Plot the graph of $y = x^2 + 2x − 5$.
   Use a scale of 2 cm to 1 unit on the x-axis and 1 cm to 1 unit on the y-axis.
(c) Use your graph to
   (i) find the value of y when $x = −1.4$.
   (ii) solve $x^2 + 2x − 5 = 0$.

6 (a) Copy and complete the table of values for $y = 8 - x^2$.

x	-3	-2	-1	0	1	2	3
8				8			8
$-x^2$				0			-9
$y = 8 - x^2$				8			-1

(b) Plot the graph of $y = 8 - x^2$.
Use a scale of 2 cm to 1 unit on the x-axis and 1 cm to 1 unit on the y-axis.

(c) Use your graph to
  (i) find the value of y when $x = 0.5$.
  (ii) solve $8 - x^2 = -2$.

7 (a) Copy and complete the table of values for $y = (x - 2)(x + 1)$.

x	-3	-2	-1	0	1	2	3	4
$x - 2$		-4					1	
$x + 1$		-1					4	
$y = (x - 2)(x + 1)$		4					4	

(b) Plot the graph of $y = (x - 2)(x + 1)$.
Use a scale of 2 cm to 1 unit on the x-axis and 1 cm to 1 unit on the y-axis.

(c) Use your graph to
  (i) find the minimum value of y.
  (ii) solve $(x - 2)(x + 1) = 2.5$.

8 (a) Make a table of values for $y = x^2 - 3x + 2$.
Choose values of x from -2 to 5.

(b) Plot the graph of $y = x^2 - 3x + 2$.
Use a scale of 2 cm to 1 unit on the x-axis and 1 cm to 1 unit on the y-axis.

(c) Use your graph to solve
  (i) $x^2 - 3x + 2 = 1$.
  (ii) $x^2 - 3x + 2 = 10$.

Exercise 11.3H

1 OABCDEFG is a cuboid.
F is the point (5, 7, 3).

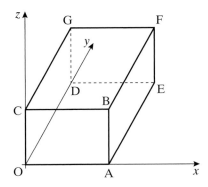

Write down the coordinates of

(a) point A.   **(b)** point B.
(c) point C.   **(d)** point D.
(e) point E.   **(f)** point G.

2 OABCV is a pyramid with a rectangular base.
V is directly above the centre of the base, N.
OA = 8 units, AB = 10 units and VN = 7 units.

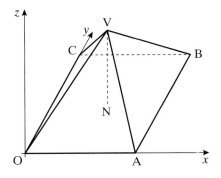

Write down the coordinates of

(a) point A.   **(b)** point B.
(c) point C.   **(d)** point N.
(e) point V.

3 OABCDEFG is a cuboid.
M is the midpoint of BF and N is the midpoint
of GF.
OA = 6 units, OC = 5 units and OD = 3 units.

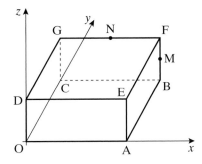

(a) Write down the coordinates of
  (i)   point B.
  (ii)   point F.
  (iii) point G.
  (iv) point M.
  (v)   point N.

(b) (i)   The point $(6, 2\frac{1}{2}, 0)$ is the midpoint of
         which edge?
   (ii)   The point $(0, 2\frac{1}{2}, 1\frac{1}{2})$ is the centre of
         which face?

Percentages

Exercise 12.1H

1 Write down the multiplier that will increase an amount by
   (a) 17%.   **(b)** 30%.   **(c)** 73%.
   (d) 6%.   **(e)** 1%.   **(f)** 12.5%.
   (g) 160%.

2 Write down the multiplier that will decrease an amount by
   (a) 13%.   **(b)** 40%.   **(c)** 35%.
   (d) 8%.   **(e)** 4%.   **(f)** 27%.
   (g) 15.5%.

3 Mrs Green bought an antique for £200.
She later sold it at 250% profit.
What did she sell it for?

4 Jane earns £14 500 per year.
She receives an increase of 2%.
Find her new salary.

5 In a sale all items are reduced by 20%.
Shamir bought a computer in the sale.
The original price was £490.
What was the sale price?

6 Graham invested £3500 at 4% compound interest.
What was the investment worth at the end of 5 years?
Give your answer to the nearest pound.

7 A car decreased in value by 11% per year.
It cost £16 500 new.
What was it worth after 4 years?
Give your answer to a sensible degree of accuracy.

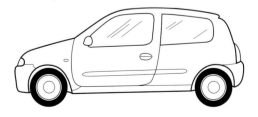

8 In a certain country the population rose by 5% every year from 2004 to 2009.
The population was 26.5 million in 2004.
What was the population in 2009?
Give your answer in millions to the nearest 0.1 of a million.

9 Jane invested £4500 with compound interest for 3 years.
She could receive either 3% interest every 6 months or 6% interest every year.
Which should Jane choose?
How much more will she receive?

10 Prices went up by 2% in 2007, 3% in 2008 and 2.5% in 2009.
An item cost £32 at the start of 2007.
What would it cost at the end of 2009?

Exercise 12.2H

 1 Wayne buys some potatoes at £1.25 a kilogram and six nectarines at 37p each.
He gives the shop assistant £10 and gets £4.68 change.
What weight of potatoes did he buy?

2 Chelsea, Sally and James share the profits from their business in the ratio 4 : 3 : 2.
In 2009 the total profit was £94 500.
Calculate how much Sally received.

3 Merry followed a recipe for lemon pudding which used 350 g of flour for four people.
He made the recipe for 10 people and used a new 1.5 kg bag of flour.
How much flour did he have left?

4 Freeville has a population of 36 281 and its area is 27.4 km².
Calculate its population density.
Give your answer to a sensible degree of accuracy.

5 Mr Brown's mobile phone bill one month showed that he had used 53 minutes of calls at 13p per minute.
The monthly rental charge for his phone was £15.30.
There was VAT at 17.5% on the whole bill.
Calculate the total bill including VAT.

6 In January 2009, the Retail Price Index (RPI) was 210.1.
In January 2010 it was 217.9.
Calculate the percentage increase in the RPI over that year.

7 In March 2005 the Average Earnings Index (AEI) was 120.0.
During the next five years it increased by 20.5%.
What was the AEI in March 2010?